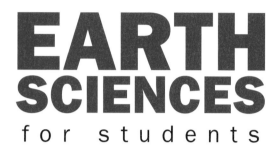

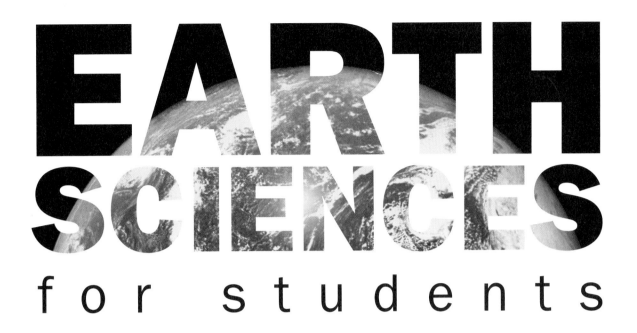

EARTH SCIENCES
for students

E. Julius Dasch

Editor in Chief

Volume

1

Macmillan Reference USA
New York

Developed for Macmillan Reference USA by Visual Education Corporation, Princeton, NJ.

Visual Education staff:
    Darryl Kestler, Editorial Director
    Theresa C. Moore, Project Director
    Amy Livingston, Project Editor
    Maureen Pancza, Copyediting Supervisor
    Maxson Crandall, Interior Design
    Cynthia C. Feldner, Electronic Preparation

Macmillan staff
    Elly Dickason, Publisher
    Hélène G. Potter, Senior Editor
    Cynthia Crippen, Indexer

Macmillan Library Reference USA
1633 Broadway
New York, NY 10019

Printed in the United States of America

printing number
1 2 3 4 5 6 7 8 9 10

Library of Congress Cataloging-in-Publication Data
Earth Sciences for Students / E. Julius Dasch, editor in chief.
        p.   cm.
    Includes bibliographical references and index.
    Summary: Describes various earth sciences, their history, employment
opportunities, and notable scientists of the past.
    ISBN 0-02-865308-4 (set). — ISBN 0-02-865309-2 (vol. 1)
    1. Earth Sciences Juvenile literature.   [1. Earth sciences.]   I. Dasch, E. Julius.
QE29.E274   1999
550—dc21   99-26905
                                                                    CIP

This paper meets the requirements of ANSI-NISO Z39.48-1991.

Preface

The earth sciences encompass not only sciences involving the solid Earth, oceanographic and atmospheric sciences, and biological sciences, but also the study of the solar system and its place in the Milky Way Galaxy and the universe. Time is a major factor that sets the earth sciences apart from most other areas of science. Because Earth's history is long—about 4.55 billion years—fundamental principles operating over such an immensity of time may be viewed in a much more evolutionary sense.

Changing Perspectives in the Field

Prior to the late eighteenth century, the roots of geology were strongly intertwined with religion, natural history, and the extraction and use of resources such as coal, metals, and building stone. The realization early in the nineteenth century that Earth's history is vastly longer than the biblical record was one of the great events of western science. The Darwinian revolution of the mid-nineteenth century, another major milestone, created a firm template for understanding paleontology and the way life has developed on Earth, and it radically redirected both geology and biology.

During the twentieth century, studies in the earth sciences shifted from descriptive and qualitative works to analytical and quantitative endeavors. In the 1940s and 1950s, geologists, physicists, and chemists began to make fundamental analytical and theoretical observations about Earth processes such as crystallization of minerals from molten rock and behavior of earthquake waves. Laboratory analysis and experimentation, combined with testable theoretical considerations, became important and necessary to the understanding of Earth. A primary example of this descriptive-to-analytical shift is the quantification of the science's most important cornerstone, the geologic timescale, which had been constructed from qualitative studies of faunal succession.

Perhaps the most dramatic breakthrough in earth sciences occurred in the 1960s and 1970s with the development of a quantitative basis for the theory of plate tectonics, which holds that large segments of Earth's lithosphere—known as plates—move about on Earth's surface relative to one another. This development provided a suitable mechanism for the drifting of continents, plate divergence (seafloor spreading), and plate collisions. Almost at once, such disparate geologic topics as earthquakes, volcanism, mountain building, and the formation and placement of ores within the Earth became understandable as parts of a grand, unified model for Earth processes.

Another basic shift in the earth sciences began in the late 1960s with the Apollo and Luna missions to the Moon, robotic exploration of many other bodies in the solar system, and the first photographs of Earth as a whole. No material better documents this shift than the first photographs from outer space of the whole Earth. These images of "Spaceship Earth" revealed the planet's wonderful colors but perilously thin atmosphere and provided people with a profoundly altered view of Earth as a unique but fragile world. Other photographs from the Apollo program and later spaceflights returned clear evidence of this fragility. Such evidence included soil erosion and deforestation from clear-cutting and the burning of large segments of the world's rain forests, a widening hole in the ozone layer at the polar regions, and ecosystems disturbed by contamination of air and water.

Missions to all of the planets except Pluto and to a large number of the satellites of the outer giant gaseous planets have documented striking and largely unforeseen differences between Earth and the other major bodies in the solar system. Extreme and unsuspected variability among the planets and their satellites has been found. Understanding this variability will help to unravel Earth's evolution.

Changing Employment Patterns and Educational Concerns

A major change in employment patterns in the earth sciences, especially in the United States and in other industrial countries, has been the marked reduction in jobs related to resource extraction and a shift to enterprises concerned more with Earth stewardship and planning. In the United States, petroleum and natural gas companies provided vast employment opportunities for geologists until the 1980s, when much lower fuel prices had a negative impact on that job market. Fortunately, opportunities in environmental areas, especially those related to water, have emerged.

Too high a percentage of the U.S. population is scientifically and technically illiterate. Addressing this matter of national concern requires better teacher preparation and in-service training, curriculum reform, better use of educational technology, and the development of national educational standards in mathematics, science, and technology. An interdisciplinary approach to learning that incorporates science, mathematics, and technology can result in a more scientifically and technically literate populace. The earth sciences provide an attractive vehicle for interdisciplinary training. Getting to know the planet Earth is a fundamental part of a balanced education.

Organization of the Material

The Macmillan *Encyclopedia of Earth Sciences* is the basis for this work. *Earth Sciences for Students* presents short discussions of various topics within the many earth science subdisciplines. Added to these topical articles is a series of entries on the history of the more fundamental subdisciplines; a similar series on employment opportunities in these major fields; and finally, 66 short biographical sketches of notable earth scientists of the past. In the original work, prominent scientists and scholars were asked to write the individual entries. The material has been reworked to clarify difficult concepts for the student, but the considerable contributions of the original authors remain reflected in the quality of this encyclopedia.

Earth Sciences for Students includes materials and devices that should aid readers in understanding complex and multidisciplinary topics. Numerous color illustrations should enhance discussion of the very visual earth sciences. Cross-references and blind entries lead the reader to other relevant topics. Essential supporting materials such as the Geologic Timescale and a Measurements and Abbreviations section are also included.

Acknowledgments and Personal Note

I wish to thank the members of the Macmillan Reference USA staff, especially Hélène Potter and Elly Dickason, for their vision and professionalism. I also wish to thank the members of Visual Education Corporation, especially Terri Moore, Amy Livingston, and the staff writers, for their painstaking editing and very effective rewriting. Mitchell Colgan, Cassandra Coombs, and Greg Coombs have my thanks for their excellent photo research. Finally, I thank my wife Patricia for many useful discussions about this work.

On a personal note—I chose geology as a career when I was about 12 years old, near the youngest age for this encyclopedia's intended readership. Perhaps the critical factor for me was a favorite aunt's employment by the Magnolia Petroleum Company (now Mobil Oil), or maybe it was my obtaining the Boy Scouts of America's merit badge "Rocks and Minerals" (now "Geology"). Whatever the reason, I never seriously considered any other profession, working first in the oil fields, then exploring for mineral deposits and ground-water resources, then university teaching, and finally government management. While geology and other earth sciences increasingly involve experimental, analytical, and computer analyses, the lure of fieldwork and travel will always remain intense, even as Sir Walter Scott noted in *St. Ronan's Well* in 1824: "and some rin up hill and down dale, knapping the chucky stanes to pieces wi' hammers, like sae mony roadmakers run daft—they sae it is to see how the warld was made!"*

E. Julius Dasch

---

*"—and some run up hill and down dale, breaking the chunky stones to pieces with hammers, like so many road workers gone mad—they say it is to see how the world was made!"

Geologic Timescale

Era	Period		Epoch	started (millions of years ago)
Cenozoic 66.4 millions of years ago–present time	**Quaternary**		Holocene	0.01
			Pleistocene	1.6
	Tertiary	**Neogene**	Pliocene	5.3
			Miocene	23.7
		Paleogene	Oligocene	36.6
			Eocene	57.8
			Paleocene	66.4
Mesozoic 245–66.4 millions of years ago	**Cretaceous**		Late	97.5
			Early	144
	Jurassic		Late	163
			Middle	187
			Early	208
	Triassic		Late	230
			Middle	240
			Early	245
Paleozoic 570–245 millions of years ago	**Permian**		Late	258
			Early	286
	Carboniferous	**Pennsylvanian**	Late	320
		Mississippian	Early	360
	Devonian		Late	374
			Middle	387
			Early	408
	Silurian		Late	421
			Early	438
	Ordovician		Late	458
			Middle	478
			Early	505
	Cambrian		Late	523
			Middle	540
			Early	570
Precambrian time 4500–570 millions of years ago				4500

Measurements and Abbreviations

The many disciplines comprising this encyclopedia—in the basic sciences, derivative studies, and applied disciplines—require an understanding of standard units of international measurements, units derived from the standard units, and "customary" (national or regional) units. The most commonly used weights and measures and their conversions are listed below.

The metric systems of units began their evolution in France in 1791. The use of these systems spread, first to non-English-speaking countries, and more recently to English-speaking countries. The first system was based on the centimeter, gram, and second (cgs) units of particular interest in science and technology. Later, the more practicable meter, kilogram, and second (mks) system was used. The present metric system is the International System of Units—Système International d'Unités, or SI, adopted in 1960. It is based on mks, and units for temperature (thermodynamic), electricity, radiation, and the quantity of substance that enters into chemical reactions have been added for a total of seven "base" units. The modern SI also includes two "supplementary" units for plane angle and solid angle, and many "derived" units for all other quantities (work, force, power) expressed in terms of the seven base and supplementary units, and other derived units. The SI metric system is now either obligatory or permissible throughout the world.

SI Base and Supplementary Unit Names and Symbols

Physical Quantity	Name	Symbol
Length	meter	m
Mass	kilogram	kg
Time	second	s
Electric current	ampere	A
Thermodynamic temperature	kelvin	K
Amount of substance	mole	mol
Luminous intensity	candela	cd
Plane angle	radian	rad
Solid angle	steradian	sr

Units Derived from SI, with Special Names and Symbols

Derived Quantity	Name of SI Unit	Symbol for SI Unit	Expression in Terms of SI Base Units	
Frequency	hertz	Hz	s-1	
Force	newton	N	m kg s-2	
Pressure, stress	Pascal	Pa	N m-2	=m-1 kg s-2
Energy, work, heat	Joule	J	N m	=m2 kg s-2
Power, radiant flux	watt	W	J s-1	=m2 kg s-3
Electric charge	coulomb	C	A s	
Electric potential, electromotive force	volt	V	J C-1	=m-2 kg s-3 A-1
Electric resistance	ohm	_	V A-1	=m2 kg s-3 A-2
Celsius temperature	degree Celsius	C	K	
Luminous flux	lumen	lm	cd sr	
Illuminance	lux	lx	cd sr m-2	

Measurements and Abbreviations

Units Used with SI, with Name, Symbol, and Values in SI Units

The following units, not part of the SI, will continue to be used in appropriate contexts (e.g., angstrom):

Physical Quantity	Name of Unit	Symbol for Unit	Value in SI Units
Time	minute	min	60 s
	hour	h	3,600 s
	day	d	86,400 s
Plane angle	degree	°	$(\pi/180)$ rad
	minute	'	$(\pi/10,800)$ rad
	second	"	$(\pi/648,000)$ rad
Length	angstrom	Å	10^{-10} m
Volume	liter	l, L	$1\ dm^3 = 10^{-3}\ m^3$
Mass	ton	t	$1\ Mg = 10^3\ kg$
	unified atomic mass unit	$u\ (= m_a(^{12}C)/12)$	$\approx 1.66054 \times 10^{-27}$ kg
Pressure	bar	bar	$10^5\ Pa = 10^5\ N\ m^{-2}$
Energy	electronvolt	$eV\ (= e \times V)$	$\approx 1.60218 \times 10^{-19}$ J

In the SI system (and, where appropriate, in SI derived units and units used together with the SI), designations of multiples and subdivisions of any unit may be arrived at by combining the name of the unit with the prefixes *deka, hecto,* and *kilo* (10, 100, and 1,000), *deci, centi,* and *milli* ($^1/_{10}$, $^1/_{100}$, $^1/_{1,000}$), and so on. These generally recognized prefixes are given here.

giga	10^9	deci	10^{-1}
mega	10^6	centi	10^{-2}
kilo	10^3	milli	10^{-3}
hecto	10^2	micro	10^{-6}
deka	10^1	nano	10^{-9}

Thus, a kilometer is 1,000 meters; a millimeter is 0.001 meter.

Conversions for Standard, Derived, and Customary Measurements

Length

1 angstrom (Å)	0.1 nanometer (exactly)
	0.000000004 inch
1 centimeter (cm)	0.3937 inches
1 foot (ft)	0.3048 meter (exactly)
1 inch (in)	2.54 centimeters (exactly)
1 kilometer (km)	0.621 mile
1 meter (m)	39.37 inches
	1.094 yards
1 mile (mi)	5,280 feet (exactly)
	1.609 kilometers
1 astronomical unit (AU)	1.495979×10^{13} cm
1 parsec (pc)	206,264.806 AU
	3.085678×10^{18} cm
	3.261633 light-years
1 light-year	9.460530×10^{17} cm

Area

1 acre	43,560 square feet (exactly)
	0.405 hectare
1 hectare	2.471 acres
1 square centimeter (cm²)	0.155 square inch
1 square foot (ft²)	929.030 square centimeters
1 square inch (in²)	6.4516 square centimeters (exactly)
1 square kilometer (km²)	247.104 acres
	0.386 square mile
1 square meter (m²)	1.196 square yards
	10.764 square feet
1 square mile (mi²)	258.999 hectares

Volume

1 barrel (bbl)*, liquid	31 to 42 gallons
1 cubic centimeter (cm³)	0.061 cubic inch
1 cubic foot (ft³)	7.481 gallons 28.316 cubic decimeters
1 cubic inch (in³)	0.554 fluid ounce
1 dram, fluid (or liquid)	$\frac{1}{8}$ fluid ounce (exactly) 0.226 cubic inch 3.697 milliliters
1 gallon (gal) (U.S.)	231 cubic inches (exactly) 3.785 liters 128 U.S. fluid ounces (exactly)
1 gallon (gal) (British Imperial)	277.42 cubic inches 1.201 U.S. gallons 4.546 liters
1 liter	1 cubic decimeter (exactly) 1.057 liquid quarts 0.908 dry quart 61.025 cubic inches
1 ounce, fluid (or liquid)	1.805 cubic inches 29.573 milliliters
1 ounce, fluid (fl oz) (British)	0.961 U.S. fluid ounce 1.734 cubic inches 28.412 milliliters
1 quart (qt), dry (U.S.)	67.201 cubic inches 1.101 liters
1 quart (qt), liquid (U.S.)	57.75 cubic inches (exactly) 0.946 liter

*There are a variety of "barrels" established by law or usage. For example, U.S. federal taxes on fermented liquors are based on a barrel of 31 gallons (141 liters); many state laws fix the "barrel for liquids" as $31\frac{1}{2}$ gallons (119.2 liters); one state fixes a 36-gallon (160.5 liters) barrel for cistern measurement; federal law recognizes a 40-gallon (178 liters) barrel for "proof spirits"; by custom, 42 gallons (159 liters) comprise a barrel of crude oil or petroleum products for statistical purposes, and this equivalent is recognized "for liquids" by four states.

Units of Mass

1 carat (ct)	200 milligrams (exactly) 3.086 grains
1 grain	64.79891 milligrams (exactly)
1 gram (g)	15.432 grains 0.035 ounce
1 kilogram (kg)	2.205 pounds
1 microgram (µg)	0.000001 gram (exactly)
1 milligram (mg)	0.015 grain
1 ounce (oz)	437.5 grains (exactly) 28.350 grams
1 pound (lb)	7,000 grains (exactly) 453.59237 grams (exactly)
1 ton, gross or long	2,240 pounds (exactly) 1.12 net tons (exactly) 1.016 metric tons
1 ton, metric (t)	2,204.623 pounds 0.984 gross ton 1.102 net tons
1 ton, net or short	2,000 pounds (exactly) 0.893 gross ton 0.907 metric ton

Pressure

1 kilogram/square centimeter (kg/cm²)	0.96784 atmosphere (atm) 14.2233 pounds/square inch (lb/in²) 0.98067 bar
1 bar	0.98692 atmosphere (atm) 1.02 kilograms/square centimeter (kg/cm²)

Measurements and Abbreviations

Energy and Power

Energy

1 joule (J)	2.390×10^{-1} calories (cal)
	9.47×10^{-4} British thermal units (Btu)
	2.78×10^{-7} kilowatt hours (kWh)
1 calorie (cal)	4.184 joule (J)
	3.986×10^{-3} British thermal units (Btu)
	1.16×10^{-6} kilowatt hours (kWh)
1 British thermal unit (Btu)	1055.87 joules (J)
	252.19 calories (cal)
	2.928×10^{-4} kilowatt hours (kWh)
1 kilowatt hour (kWh)	3.6×10^{6} joules (J)
	8.60×10^{5} calories (cal)
	3.41×10^{3} British thermal units (Btu)

Power (energy per unit time)

1 watt (W)	3.4129 British thermal units per hour (Btu/h)
	1.341×10^{-3} horsepower (hp)
	14.34 calories per minute (cal/min)
1 horsepower (hp)	550 ft lb/s
	7.46×10^{2} watts (W)

Temperature

Scientists commonly use the Celsius system. Although not recommended for scientific and technical use, earth scientists also use the familiar Fahrenheit temperature scale (°F). 1°F = 1.8°C or K. The triple point of H_2O, where gas, liquid, and solid water coexist, is 32°F.

- To change from Fahrenheit (F) to Celsius (C):
  °C = (°F-32)/(1.8)
- To change from Celsius (C) to Fahrenheit (F):
  °F = (°C × 1.8) + 32
- To change from Celsius (C) to Kelvin (K):
  K = °C + 273.15
- To change from Fahrenheit (F) to Kelvin (K):
  K = (°F-32)/1.8 + 273.15

Abbreviations

AABW: Antarctic Bottom Water

AAG: Association of American Geographers

AAIW: Antarctic Intermediate Water

ABET: Accreditation Board for Engineers and Technologists

ACCP: Atlantic Climate Change Program

ADEOS: Advanced Earth Observing System

AEG: Association of Engineering Geologists

AGI: American Geological Institute

API: American Petroleum Institute

ASCE: American Society of Civil Engineers

ATLAS: Atmosphere Laboratory for Applications and Science

ATOC: Acoustic Thermometry for Ocean Climate

AU: astronomical units

Bbo: barrels of oil

BBO: billion barrels of oil

BOD: biological oxygen demand

CD-ROM: compact disc read-only memory

CMB: core-mantle boundary

CME: coronal mass ejection

COCORP: Consortium for Continental Reflection Profiling

CT: carat

D/H: deuterium-to-hydrogen ratio

DSDP: Deep Sea Drilling Project

DTA: Differential Thermal Analysis

EELS: electron energy loss spectroscopy

EOS: Earth Observing System

EOSDIS: EOS Data and Information System

EPA: Environmental Protection Agency

ERBE: Earth Radiation Budget Experiment

ESA: European Space Agency

FAMOUS: French-American Mid-Ocean Undersea Study

Ga: billion years

GCM: General Circulation Model

GIS: geographical information system

GPS: Global Positioning System

GSA: Geographical Society of America

GWP: Global Warning Potential

ICB: inner core boundary

ICES: International Council for the Exploration of the Sea

IGY: International Geophysical Year

IPCC: Intergovernmental Panel on Climate Change

JOI: Joint Oceanographic Institutions

JOIDES: Joint Oceanographic Institutions for Deep Earth Sampling

Ka: thousand years

LAGEOS: Laser Geodynamics Satellite

LANDSAT: Land Remote-Sensing Satellite

LDG: Libyan Desert Glass

LITE: Lidar In-Space Technology Experiment

Ma: million years

MESUR: Mars Environmental Survey Mission

MORB: mid-ocean ridge basalt

MSF: multistage flash

NADW: North Atlantic Deep Water

NAS: National Academy of Sciences

NASA: National Aeronautics and Space Administration

NEOs: near-Earth objects

NMR: nuclear magnetic resonance

NOAA: National Oceanic and Atmospheric Administration

NPC: National Petroleum Council

NSF: National Science Foundation

NTIS: National Technical Information Services

NWS: National Weather Service

ODP: Ocean Drilling Project

PCE: pyrometric cone equivalent

PDR: precision depth recorder

PGE: platinum-group elements

PICs: products of incomplete combustion

PIXE: proton-induced X-ray emission

ppb: parts per billion

ppm: parts per million

ppt: parts per thousand

REEs: rare-earth elements

RO: reverse osmosis

ROV: remotely operated vehicle

SAR: Synthetic Aperture Radar

SeaWiFS: Sea-Viewing Wide Field Sensor

SEM: scanning electron microscope

SETI: Search for Extraterrestrial Intelligence

SLR: satellite laser ranging

SNC: shergottite, nakhlite, and chassignite

SOC: synthetic organic chemical

SST: sea surface topography

TDS: total dissolved solids

TMDL: total maximum daily load

UNEP: United Nations Environment Program

UNESCO: United Nations Educational, Scientific, and Cultural Organization

USGS: U.S. Geological Survey

VLBI: Very Long Baseline Interferometry

VOC: volatile organic chemical

WOCE: World Ocean Circulation Experiment

WWSSN: World Wide Standardized Seismograph Network

WWW: World Wide Web

List of Contributors

The text of *Earth Sciences for Students* is based on the Macmillan *Encyclopedia of Earth Sciences*, which was published in 1996. We have updated the material where necessary and added a few new entries. Here we wish to acknowledge the authors of the original articles and the new entries:

Dallas H. Abbott
Allan Allwardt
Don Anderson
Phillip Anz-Meador
Bob Arnold
James Arnold
Ghassem Asrar
David H. Atkinson
Jayne Aubele
Loren Babcock
William Back
Jean M. Bahr
Arthur Barber
Nadine G. Barlow
Eric Barron
Raymond Batson
James S. Beard
Stephen Becker
Stefan Bengston
Keith R. Benson
William B. N. Berry
Jacob Bigeleisen
Gordon Bjoraker
David C. Black
Geoffrey Blewitt
Arthur Bloom
Sam Boggs
Kennard Bork
Harold W. Borns
Arthur Boucot
Joseph Boyce
Danita Brandt
Ron Brashear
Carlton E. Brett
Geoffrey Briggs
Richard G. Bromley
Neal Brown
Raymon L. Brown
Bonnie Buratti
Robert W. Burpee

Louis J. Cabri
Humberto Campins
Philip A. Candela
Robert L. Carroll
Rufus D. Catchings
Clark R. Chapman
Alan J. Charig
Nikolas Christensen
Mark Cintala
Faruk Civan
Donald Coates
Laurel Collins
Tim Collins
Vickie S. Connors
Catherine Constable
Gordon Craig
James R. Craig
Dale P. Cruikshank
Larry S. Crumpler
Cindy Cunningham
Scott Curry
Heinz Damberger
E. Julius Dasch
Patricia Dasch
John Degnan
Richard V. Dietrich
Donald B. Dingwell
Deborah Domingue
Ellen Tan Drake
Larry Drew
Linda E. Duguay
John R. Dyni
Adam M. Dziewonski
Farouk El-Baz
James L. Elliot
Donald Elthon
Lawrence Engel
Terry Engelder
W. Gary Ernst
Doug Erwin

Eric J. Essene
Frederick C. Eubank
Ronald D. Evans
Paul Farber
Gunter Faure
P. Geoffery Feiss
Cyrus W. Field
D. Chris Findlay
William L. Fisher
James R. Fleming
Kenneth A. Foland
Robert L. Folk
Eric R. Force
Clyde Frank
Gerald M. Friedman
Louis Friedman
Robert Marc Friedman
William S. Fyfe
Patricia Gensel
Matthew P. Golombek
Pamela Gore
Joseph L. Graf
Jane Gray
David H. Green
Mott Greene
M. Grant Gross
Virginia C. Gulick
Anuj Gupta
Bradford H. Hager
Anthony Hallam
Mark D. Hannington
Candice Hansen
Dena Hanson
Charles R. Harper
William K. Hartmann
Allen Hatheway
B. Ray Hawke
Peter Heaney
G. Ross Heath
Thomas A. Herring

Peter M. Herzig
Carole Hickman
Mark L. Hineline
Erik Hjortenberg
E. Dorrit Hoffleit
Paul F. Hoffman
Deidre Hunter
Robert Hutchison
Philip Dean Ihinger
Andrew Ingersoll
Raymond Jeanloz
Amy J. G. Jurewicz
Stephen R. Jurewicz
Hiroo Kanamori
Ellen Kappel
Stephanie Kehoe
Scott King
Cornelis Klein
Christian Koeberl
James B. Koenig
Paul Komar
Frederick Koomanoff
Konrad Krauskopf
Keith A. Kvenvolden
Gene L. LaBerge
Willard C. Lacy
Nicholas Lancaster
Ian M. Lange
Elana Leithold
David Levy
John M. Lewis
Robert J. Lillie
Jere H. Lipps
Robert Livingston
David Long
John A. Long
Rosaly Lopes-Gautier
John R. Lukacs
James Luyten
Kirk A. Maasch
Renu Malhotra
Warren Manspeizer
Gene Mapes
Robert L. Marcialis
Mark Marley
Bruce Marsh
Susan Marshall
Ursula B. Marvin
Brian Mason
Christopher C.
  Mathewson
Ted A. Maxwell
J. Barry Maynard
Alexander R. McBirney

Les McFadden
Brian McGowran
Kenneth C. McGwire
Marcia McNutt
Michael Meyer
Richard F. Meyer
Kitty Milliken
Kula C. Misra
David W. Mittlefehldt
George W. Moore
John E. Moore
S. Conway Morris
Richard A. Muller
Haydn H. Murray
John Mutter
Clifford M. Nelson
Dennis O. Nelson
Horton E. Newsom
Elisabeth G. Newton
Lawrence E. Nyquist
James O'Donnell
Gerhard Oertel
Jack E. Oliver
Peter Olson
Naomi Oreskes
Nathaniel Ostrom
Virginia M. Oversby
Peter Patton
W. Richard Peltier
Robert A. Phinney
Jonathan G. Price
Kurt Putnam
Jeff Raffensperger
Ronald Rainger
George (Rip) Rapp
Greg Retallack
Winifred Reuning
Paul Roberts
Eugene C. Robertson
Michael G. Rochester
Chris S. Romanek
John Ruben
Constance Sancetta
Richard Sanford
Dale Sawyer
Charles K. Scharnberger
Harrison Schmitt
Jill S. Schneiderman
Robert Schoch
Wilfried Schroder
Christopher R. Scotese
Martha Scott
Keith Seitter
Stephen Self

Jane Selverstone
George H. Shaw
Stephanie S. Shipp
Dave Shutler
Brian J. Skinner
H. Catherine W. Skinner
Judy Skog
Lisa Sloan
Phillip Sloan
Raymond N. Smartt
Scott Southworth
Frank S. Spear
Frank D. Stacey
Ronald W. Stanton
Eugen F. Stumpfl
Marilyn J. Suiter
Artur Svansson
Ralph E. Taggart
David Talent
Lynne Talley
Manik Talwani
David Taylor
Peter Thomas
Spencer R. Titley
Allen Treiman
James R. Underwood
Thomas M. Usselman
Christopher
  VanCantfort
David R. Veblen
Michael Velbel
John D. Vitek
Garth Voigt
James C. G. Walker
William White
Michael Whiteley
Michael L. Wiggins
Jack A. Williams
Richard S. Williams, Jr.
Stanley N. Williams
Steven H. Williams
Leonard Wilson
Gary R. Winkler
Ellen Wohl
M. Gordon Wolman
Lynn Wright
Robert S. Yeats
Ellis Yochelson
Hatten S. Yoder
Half Zantop
Mary Lou Zoback
Michael E. Zolensky

Table of Contents

Volume 1

Volume 2

Table of Contents

Volume 3

Volume 4

Abrasive Materials

Abrasive materials are substances used to abrade, or actively wear away, surfaces. There is a wide variety of both natural and manufactured abrasives used for many purposes, from scouring industrial steel to polishing teeth.

Abrasive materials usually occur in the form of particles. They may be used in grit or powder form, in slurries (mixtures of abrasive material with water or oil), or in pastes such as toothpaste and jewelry polish. Abrasives may be embedded in or fastened to metal, such as saw blades or knife sharpeners.

Types of Abrasives. Many abrasives occur naturally; these range in particle size from huge boulders to fine grit. Natural abrasion takes place on beaches and in streambeds, where the constant movement of the water rubs rocks against each other, making them round and smooth. The same natural process occurs when wind, water, or ice (such as a glacier) moves materials such as sand, grit, and boulders, producing wind-carved rock formations, glacial grooves, and other landscape features.

One of the most widely used natural abrasives is diamond, the hardest known substance. Others include garnet, quartz, pumice, and sand. Emery boards and emery wheels are covered with an abrasive made of crushed "emery" rock, which contains iron oxides and the mineral corundum. Tripoli, a category of "soft" abrasives, consists of grains that have no sharp corners or edges. It is made by crushing weathered rocks that are rich in silica.

Manufactured abrasives include glass, finely powdered iron oxide (called "rouge"), and metallic abrasives such as steel wool. Synthetic, or manufactured, diamonds have largely replaced natural diamonds as abrasives because their properties or shapes can be tailored to specific industrial uses.

Uses of Abrasives. Abrasives can be used either to roughen or to smooth surfaces. The effect depends on the size and hardness of the abrasive particles and on the force used, from gentle rubbing to heavy-duty blasting with air or water. Larger and harder abrasives grind surfaces into shapes and remove large fragments or bumps, but they also create small scratches and a rough texture. Finer and softer abrasives remove the scratches.

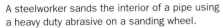

A steelworker sands the interior of a pipe using a heavy duty abrasive on a sanding wheel.

One of the main industrial uses of abrasive grit is in work with steel and other metals. Manufacturers, craftspeople, and hobbyists also use abrasive materials to finish wood, leather, and plastic. They buff varnished or painted surfaces, polish gemstones, sharpen tools, and create high-quality, scratch-free surfaces on the semiconductors used in computers. Abrasives are also used to grind and polish glass, including eyeglass lenses. Nonskid strips on ramps and stair treads or in bathtubs are coated with abrasive material to give them their "grippy" quality. Tripoli and other mild abrasives are used in scouring powders, toothpastes, and buffers for fine objects such as ceramics and pearls.

The widespread use of abrasive materials in everything from common household products to industrial grinding and polishing devices makes the manufacture and sale of these materials a very big business. In 1989 alone, U.S. production of abrasive materials accounted for $227,761,000; exports accounted for $260,363,000.

Absolute Dating

See Geologic Time.

Acid Rain

See Pollution of the Atmosphere, Pollution of Streams, Lakes, and Groundwater; Weathering and Erosion.

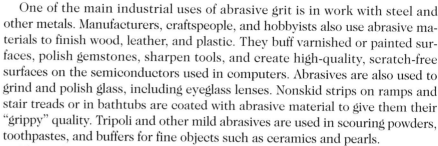

Agassiz, Jean Louis Rodolphe

1807–1873
Swiss naturalist

natural history   systematic study of natural objects, especially in their natural settings (includes physical and life sciences)

paleontology   science that deals with prehistoric plants and animals through the study of fossils and other remains

comparative anatomist   scientist who compares the structures of different plants and animals

vertebrate   any animal that has a backbone or spinal column

Jean Louis Rodolphe Agassiz, a leading figure in **natural history** during the 1800s, spent much of his working life in the United States. Agassiz's broad interests included biology, **paleontology**, and geology. He is best remembered for his study of glaciers—his work led to the idea of the "Ice Age"—and for his opposition to Charles Darwin's theory of evolution.

Fossil Fishes and Moving Ice.   Agassiz was born in Motiers, Switzerland. After receiving a medical degree in 1830 from the University of Heidelberg, Germany, he returned his attention to his lifelong interest, natural history. Agassiz went to Paris to study under the great **comparative anatomist** and founder of **vertebrate** paleontology, Georges Cuvier.

In 1832 Agassiz became a professor of natural history at the College of Neuchâtel in Switzerland, where he remained until 1846. During those years, he launched his scientific career by publishing a massive five-volume study of fossil fishes as well as other works on fossils and biology.

During this time, Agassiz also became interested in glacial geology, the study of glaciers and how they shape the Earth's surface. The Swiss Alps provided a perfect outdoor laboratory. In one experiment, Agassiz drove a straight line of stakes into the ice and returned years later to find that the middle stakes had moved downhill, helping him to prove that glaciers gradually move or flow. He also noted that glaciers leave behind distinctive markings and piles of rock when they travel over land. The existence of similar features in northern Europe and North America suggested to him that these regions had once been covered with large ice sheets. By helping to establish the concept of an ice age, Agassiz broadened our understanding of the Earth and its history.

An American Scientist.   In 1846 Agassiz gave a series of lectures in Boston. The success of his lectures, plus his interest in the natural history of the United States, resulted in a permanent move to America. In 1848 he became a professor

Jean Louis Rodolphe Agassiz

at Harvard University in Cambridge, Massachusetts, which remained his home base until his death.

Agassiz earned a reputation as an enthusiastic teacher and lecturer. Well known in scientific circles, he traveled widely and published a number of books on biological and geological topics. Agassiz founded Harvard's Museum of Comparative Zoology in 1859, the same year in which he declined a professorship at the Museum of Natural History in Paris. In 1863 he helped to found the National Academy of Sciences in Washington, D.C.

Throughout his career, Agassiz opposed the theory of evolution that Charles Darwin and other scientists had proposed to explain how life-forms changed over time. Agassiz knew that differing fossil forms outline the history of life: his fossil studies helped chart that history. However, he believed that natural disasters had wiped out each vanished stage of life and that the new, more advanced organisms that emerged after each disaster were newly created as well as fixed and incapable of change.

Six years after Agassiz's death, scientists named an ancient lake, which had once covered parts of Minnesota, Manitoba, and North Dakota, Lake Agassiz in his honor. He was elected to the Hall of Fame for Great Americans in 1915. SEE ALSO CUVIER, GEORGES; DARWIN, CHARLES ROBERT; EARTH AS A DYNAMIC SYSTEM; GLACIAL AGES; LIFE, EVOLUTION OF.

Alloy Metals

An alloy is a substance formed by combining one type of metal with other metals or nonmetals. Common examples of alloys include brass, which is a mixture of copper and zinc, and stainless steel, which is a combination of iron, carbon, and chromium. Alloys are used to make countless items from household utensils to spacecraft.

Characteristics of Alloys. Pure metals have certain measurable physical properties, including hardness, strength, resistance to corrosion (such as rust), and ability to conduct heat and electricity. Alloys can have physical properties that differ significantly from those of the pure metals from which they are made. In fact, it is possible to create an alloy that has just the right combination of properties for a particular use. That is why most metallic materials in use today are alloys.

There are two main categories of alloys: ferrous alloys, which contain significant amounts of iron, and nonferrous alloys, which consist primarily or entirely of substances other than iron. Alloys have three main types of internal structure. Some alloys are chemical mixtures of metals that blend completely, resulting in a uniform internal structure very much like that of the main ingredient, or base metal. In other alloys, the metals do not mix completely with each other, and two or more distinct internal structures (still similar to the base metal's) result. In a third type of alloy, the metals combine in the precise proportions of a chemical compound and form a new internal structure, very different from those of the original metals.

Compounds of Alloys. Among the most important alloys is steel, a combination of iron and carbon. A number of metals—including chromium, molybdenum, tungsten, nickel, cobalt, and manganese—are added to steel to produce certain physical and chemical properties for specific purposes. For example, stainless steel—steel that contains at least 11 percent chromium—is strong, hard, and resistant to heat and corrosion.

A worker in a bell factory pours molten alloy into a mold.

The metals and nonmetals used to make alloys are found in scattered deposits of ore throughout the Earth's crust. Some are more plentiful and easier to obtain than others. Like other materials in the Earth's crust, the metal ores were formed by geologic processes over the course of millions or billions of years, and their quantities and locations depend on various geologic factors. The world's largest chromium deposits, for example, are located in South Africa, whereas one of the greatest concentrations of tungsten deposits is in China.

New alloys and new technologies go hand in hand. Alloys have made the possibility of space travel a reality, have furthered the exploration of the oceans to their deepest depths, and have allowed for the design of computers the size of a notebook. Scientists and engineers continue to develop new alloys for the technologies of the future. SEE ALSO CHROMIUM; EARTH MATERIALS, CHEMISTRY OF; MANGANESE DEPOSITS; METALS, DISTRIBUTION OF; MINERAL DEPOSITS, FORMATION OF; MINERAL DEPOSITS, METALLIC; TITANIUM DEPOSITS.

Alluvial Plain

See Rivers, Erosion and Deposition by.

Aluminum Resources

Aluminum is a silvery white metal that is strong but light, resistant to rust, a good conductor of electricity, and geologically plentiful. Because of these characteristics, aluminum is one of the most important resources in industry. It is used in construction and in the manufacture of cars and airplanes, electrical equipment, machinery, beverage cans and other containers, and many household products.

Aluminum is the most abundant metal and the third most abundant chemical element in the Earth's crust. It is a major component of many common minerals. Despite its abundance, aluminum is difficult to separate from the elements with which it combines in nature. Aluminum metal was first produced in a pure form only in 1827, and the separation process was so expensive at that time that the metal was used only to make jewelry and other costly items.

Aluminum metal first became available in large amounts and at reasonable prices in 1886, following the development of a new method of separating the metal from ore. The separation process used today, based partially on that earlier process, produces a molten metal that can be formed into bars, or ingots, of pure aluminum. Aluminum production requires large amounts of heat and electricity.

Although a large number of minerals contain aluminum, only a few are desirable ores—ores from which it is relatively easy and cheap to separate the metal. The most important aluminum ore is bauxite, a mineral formed by intense weathering that most commonly occurs in tropical regions. Although the United States has no major bauxite deposits, there are some bauxite reserves in central Arkansas. It is estimated that worldwide resources of bauxite will last well into the 2000s. Eventually, however, new sources of aluminum will have to be found. SEE ALSO INDUSTRIAL MINERALS; METALS, DISTRIBUTION OF; MINERAL DEPOSITS, METALLIC; MINERALS; MINERAL SUBSTANCES, USEFUL; NONFERROUS METALS.

Alvarez, Luis Walter

1911–1988
American physicist

Luis Walter Alvarez.

Luis Walter Alvarez was an American physicist, Nobel laureate, and inventor who, thanks to his son Walter, made a major discovery in geology and astronomy. Alvarez was born in San Francisco in 1911. He studied physics at the University of Chicago; then he went to the University of California at Berkeley to work for Ernest Lawrence, who had recently invented and built a new kind of atom smasher called a cyclotron. In the next few years at Berkeley, Alvarez used the new machine to make many important discoveries about the nucleus (plural *nuclei*), or center, of an atom.

When World War II broke out, Alvarez used his expertise to help the military. He invented a radar method to help airplanes land even in zero visibility. Later in the war, he worked on the atomic bomb project and invented a way of measuring the energy released by the bomb. Traveling in an airplane over Hiroshima as an atomic bomb was dropped by another plane, Alvarez determined that the bomb's energy equaled that of 13,000 of TNT.

After the war, Alvarez returned to Berkeley, where he built a new device called the "Liquid Hydrogen Bubble Chamber" for studying atomic nuclei. This invention allowed him to make images of the particles coming from a nuclear collision, and it led directly to the discovery of a large number of subnuclear particles. For this work, Alvarez was awarded the Nobel Prize in physics in 1968.

Alvarez was known for his imagination and willingness to attempt daring experiments, such as his use of cosmic rays to "X-ray" the pyramids of Egypt. That experiment was a success, and the image was obtained, although no new chambers were revealed.

Perhaps Alvarez will be best remembered for his analysis of the great biological catastrophe that took place about 66 million years ago, in which the dinosaurs and most other life-forms disappeared. Alvarez had never before worked in geology; he became interested in the problem through his son Walter, a geologist. Working together, they discovered, in several locations worldwide, a thin layer of clay that had been deposited about 66 million years ago. This clay is enriched in the element iridium—a clear chemical indicator, or sign, of debris from the impact of a comet or asteroid. Years later the crater formed by the impact was located on the Yucatán Peninsula in Mexico. This discovery changed many scientists' views about evolution: Species must survive not only competition with other species but also an occasional catastrophic event if they are to endure over millions of years.

Amphibians and Reptiles

paleontologist   scientist who studies prehistoric plants and animals through fossils and other remains

species   narrowest classification, or grouping, of organisms according to their characteristics; members of a species can reproduce only with others of that group

Amphibians and reptiles are two categories of living animals whose ancestors first appeared on Earth hundreds of millions of years ago. **Paleontologists** study fossil remains to learn how present-day reptiles and amphibians descended from past forms of these animals; how and when reptiles and amphibians began to evolve as separate **species**; and how other modern species developed—all parts of the complex history of life on Earth. A tadpole swimming in a pond and a snake gliding across a rock are not only fascinating creatures but also living links with the prehistoric world.

Amphibians

Animals that live partly in the water and partly on land are called amphibians. The three main groups of amphibians are frogs, salamanders, and legless creatures called caecilians that live in the tropics. Most amphibians lay their eggs in water.

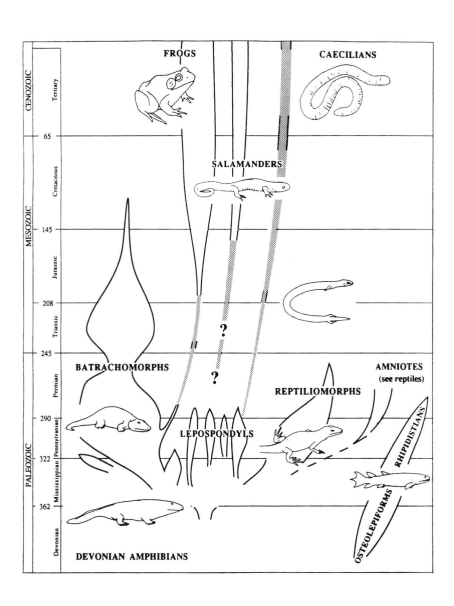

Geological distribution of the major groups of fossils and living amphibians. Numerical scale on the left is in millions of years.

GT See Geologic Timescale on page viii.

Early Forms. The earliest known amphibian fossils come from the Devonian period^GT, more than 360 million years ago. The skeletons of these early amphibians have features of both fish and primitive reptiles. Scientists call the first amphibians stegocephalians, which means "roofed skulls," because they had solid bony coverings over their heads. Stegocephalians were the dominant land animals for 70 million years.

Two types of amphibians emerged during the Paleozoic era: the labyrinthodonts and the lepospondyls. The labyrinthodonts were large—generally more than a meter long—and lived in a wide variety of habitats. Scientists believe that all modern reptiles, birds, and mammals are descendants of a subcategory of labyrinthodonts called Reptiliomorpha. Another subcategory, Batrachomorpha, probably included the ancestors of some present-day amphibians. The lepospondyls died out before the Mesozoic era began, over 245 million years ago.

Origin of Modern Forms. The earliest fossils that resemble modern amphibians date to the Jurassic period, 208 to 144 million years ago. Jurassic frogs and salamanders are similar to those seen today, but Jurassic caecilians still had tiny legs. All modern families of frogs, salamanders, and caecilians had appeared by the beginning of the Cenozoic era, 66 million years ago.

New Zealand's Living Fossils

A group of reptiles called Rhynchocephalia, or "beak-headed," appeared about 220 million years ago, before the dinosaurs. Rhynchocephalia became extinct about 100 million years ago—all but one type, known as Sphenodon. On isolated islands off the coast of New Zealand in the Pacific Ocean, these small, lizard-like reptiles have survived to the present day. Called tuataras by New Zealanders, they may live as long as 100 years. The tuatara's eggs incubate for longer than those of any other reptile—up to 15 months before they hatch.

Reptiles

Some of the most spectacular fossils are those of the dinosaurs, flying reptiles, and giant water-dwelling reptiles that dominated the land, air, and sea during the Mesozoic era. The reptile groups that survive today are lizards, snakes, and turtles. Like birds and mammals, reptiles are amniotes, meaning that their embryos are surrounded by membranes that allow the eggs to be laid on land or held inside the mother's body.

Early Amniotes. Fossils of amniotes first appear in rocks that are 320 to 286 million years old. One group, the Synapsida, had traits of both reptiles and early mammals and eventually led to the development of mammals. Although the word *synapsid* means "mammal-like reptile," most experts now group these animals with mammals rather than with reptiles.

The second group of early amniotes produced reptiles and birds. It split into two divisions, the Chelonia, or turtles, and the Diapsida. The first fossils of turtles are more than 210 million years old. They had heavy protective shells but could not draw their heads inside the shells as modern turtles do. The process of development that led from primitive reptiles to the Chelonia is not well understood, and scientists have continued to debate turtles' ancestry.

Many varieties of diapsids appeared during the late Paleozoic era. Around 250 million years ago, a group called the lepidosauromorphs produced the ancestors of snakes, lizards, and a group of reptiles called Sphenodons. Although many species of snakes and lizards exist today, only one species of Sphenodon has survived.

Another group of diapsids, the archosauromorphs, includes dinosaurs, pterosaurs (flying reptiles), crocodiles, and birds.

These animals could draw their legs more directly under their bodies than the lepidosauromorphs could. This structural variation enabled some dinosaurs, and later the birds, to stand erect on two feet.

Amphibians and Reptiles

Geological distribution of the major groups of fossils and living reptiles. Numerical scale on the left is in millions of years.

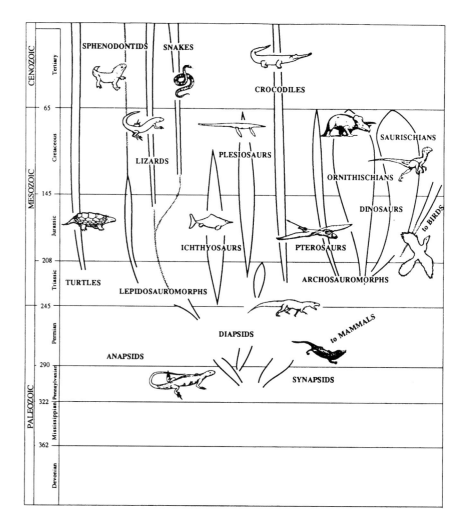

carnivorous meat-eating
herbivorous plant-eating

The Age of Reptiles.   Among the reptiles that appeared in the Triassic period, from 245 to 208 million years ago, were long-legged crocodiles that lived on land. Over time some varieties of crocodiles adapted to living in the water.

Large-winged pterosaurs soared through the skies during this period. Below them lived the **carnivorous** and **herbivorous** dinosaurs. By the late 1900s, paleontologists had determined that birds had evolved from small dinosaurs early in the Mesozoic era. However, several fossils discovered in the 1990s—including the earlier known beaked bird and several feathered dinosaurs—indicate that scientists have not yet fully unraveled the complex relationship between dinosaurs and birds.

Also living in the Triassic period were several groups of marine reptiles. The icthyosaurs could swim rapidly; some species were shaped much like the tuna, the fastest modern fish. Some of a group known as placodonts evolved a bony covering similar to a turtle's shell. The plesiosaurs, some of which were more than 12 m long, had limbs shaped like giant paddles that propelled them through the water like the flippers of modern seals.

Dinosaurs, pterosaurs, and marine reptiles were very numerous until the end of the Cretaceous period, around 66 million years ago, when they suddenly became extinct. The ancestors of snakes, lizards, and turtles survived to produce the species we know today. **SEE ALSO** Dinosaurs; Extinctions.

Ancient Supercontinents

If the Earth's continents could be moved around like pieces of an enormous jigsaw puzzle, the eastern side of South America and the western side of Africa would fit neatly together. From 300 to 200 million years ago, in fact, they did fit together as part of a single large supercontinent. This landmass eventually split into two supercontinents, from which the continents we know today were formed. By the late 1900s, scientists came to believe that the continents repeatedly came together and fragmented in the course of Earth's history.

Moving Continents.   In the early 1900s, a German scientist named Alfred Wegener argued that all the continents had once been joined together in what he called an ancient supercontinent. He named it Pangaea (from Greek words meaning "all earth").

Wegener had noticed that the continents appeared to fit together like giant puzzle pieces—and that, when they were fitted together, geological features such as mountain ranges matched up from one continent to the next. Other evidence seemed to support Wegener's claim that the continents had not always been in their present position. Moving continents could explain the presence of coral reef fossils in the Arctic and traces of glaciers in the Sahara. Another hint that the landmasses had once been connected was the fact that plants and animals that could not cross oceans lived on widely separated continents.

Wegener's theory of phenomenon called continental drift explained some geological mysteries, but no one could explain just how the continents were supposed to have moved. In the second half of the 1900s, however, new scientific discoveries gave rise to a theory, called plate tectonics, that offers an explanation for how the continents have drifted slowly over millions of years. **Convection currents** within the hot **mantle** appear to carry the less dense continents over Earth's surface much as if they were pieces of wood floating on water.

Geography of the Ancient World.   Scientists have come to agree that Wegener was right about the ancient supercontinent. Geologists have been able to plot the past paths of the continents, using a technique called paleomagnetism. By measuring the traces that the Earth's magnetic field has left in iron-bearing rock, they can tell where on the Earth's surface such rock was formed; by measuring variations in the magnetic field of the Earth's oceans, they can reconstruct past shapes of the ocean basins. Paleomagnetism has allowed scientists to see how the continents have moved over the past 200 million years and, less clearly, how they may have moved before that time.

For 100 million years or more, the Earth had a single landmass, Pangaea, formed by the collision of several continents. Around 150 million years ago, the forces of plate tectonics caused Pangaea to break slowly apart into two supercontinents. The northern one, Laurasia, included what is now North America, Europe, and parts of Asia. The southern one, Gondwana or Gondwanaland, included South America, Africa, India, China, Australia, and Antarctica.

Plate tectonics continued to shape the ocean basins and shift the continents, pulling Laurasia and Gondwana apart and moving the pieces into their present positions. North America separated from Eurasia; then South America separated from Gondwana. India broke loose from Gondwana and drifted north, colliding with Eurasia about 50 million years ago. The Himalaya, the world's highest mountain range, marks the site of that slow but immensely powerful collision.

The plates that carry the Earth's landmasses and ocean basins are still drifting, at speeds ranging from 2 to 10 cm a year. In another hundred million years, the continents and oceans of the modern globe may be as much a part of the past as Pangaea, Laurasia, and Gondwana. SEE ALSO MAGNETIC FIELD, EARTH'S; PLATE TECTONICS; WEGENER, ALFRED LOTHAR.

convection current   movement of a fluid (liquid or gas) caused when denser portions of the fluid move downward because of gravitational attraction, pushing less dense portions up and away

mantle   region of the Earth between the molten core and the outer crust

Angiosperms

Angiosperms, or flowering plants, have their seeds enclosed within fruit. Comprising about 300,000 different **species,** angiosperms are the dominant form of plant life on Earth. They include all grains and grasses, all flowers, many common broad-leaved shrubs and trees, and most weeds. Angiosperms are found everywhere on Earth except in the oceans and at the Antarctic ice cap.

Angiosperms developed more than 135 million years ago during the Cretaceous period[GT] of the Mesozoic era. Some scientists believe that the ancestor of modern angiosperms was a woody shrub with large flowers. Others are convinced that the ancestral angiosperm was a small-flowered, herbaceous (nonwoody) plant. Angiosperms rapidly developed a variety of successful forms during the early Cretaceous period. First appearing along riverbanks in regions near the equator, they spread to many other environments. Their spread accompanied the **extinction** of a number of other plant groups and dramatically changed the Earth's vegetation.

Fossils of ancient angiosperms have helped scientists determine what the climate was like in various regions on Earth millions of years ago. In addition, by using dating techniques to establish the age of rocks that contain angiosperm fossils, scientists can estimate the age of strata, or layers, of the Earth's surface where the fossils are located elsewhere. Angiosperm fossils have also been helpful in testing theories about the distribution of plant and animal species over time. SEE ALSO FOSSILIZATION AND THE FOSSIL RECORD; GEOLOGIC TIME; PALEO-BOTANY, PALEOCLIMATOLOGY, PALEOECOLOGY, PALEOGEOGRAPHY.

Apollo Astronauts

The goal of the Apollo space program was to put astronauts on the Moon. The Apollo missions began in 1967, met the goal, and laid the foundation for future space exploration.

Astronauts in the Apollo program included experienced military pilots, scientists, and engineers. Among them were seven men who had participated in Project Mercury, the program that sent the first American into space in 1961 and the first American into orbit around the Earth in 1962.

Of the 73 men selected for the Apollo program, 24 traveled to the Moon; of those, 12 set foot there. The other astronauts contributed expertise and in-

Astronaut Buzz Aldrin, lunar module pilot, stands beside an American flag placed on the Moon during the Apollo 11 mission in 1969. He is photographed by the mission's commanding officer, Neil Armstrong.

sight, oversaw engineering projects, and participated in all aspects of mission support.

Training of the Apollo astronauts concentrated on improving their observational and data-gathering skills. They also received geological field training so that they would be able to understand and gather information about the geology of the Moon.

There were 17 Apollo missions between 1967 and 1972; of these, 6 landed on the Moon. These missions took humankind into the solar system for the first time and led to a greater understanding of the evolution, structure, and composition of the Moon as well as of the Earth. The Apollo program also established the psychological, technological, and scientific foundation on which to base the planning of future space missions. SEE ALSO MOON; SOLAR SYSTEM; SPACE EXPLORATION.

Manned Apollo Missions

Flights	Astronauts	Mission
Apollo I 1/27/57	Virgil I. Grissom Edward H. White II Roger B. Chaffee	Crew killed during test launch on pad
Apollo VII 10/11/68	Walter M. Schirra R. Walter Cunningham Donn F. Eisele	Earth orbit test flight of the Command and Service Module
Apollo VIII 12/21/68	Frank Borman James A. Lovell William A. Anders	Lunar orbit test flight of the Command and Service Module
Apollo IX 3/3/69	James A. McDivitt David R. Scott Russell L. Schweikart	Earth orbit test flight of the Command and Service Module and Lunar Module
Apollo X 5/18/69	Thomas P. Stafford John W. Young Eugene A. Cernan	Lunar orbit test flight of the Command and Service Module and Lunar Module
Apollo XI 7/16/69	*Neil A. Armstrong Michael Collins *Buzz Aldrin	First lunar landing, Mare Tranquilitatis
Apollo XII 11/14/69	*Charles Conrad, Jr. Richard F. Gordon, Jr. *Alan L. Bean	Second lunar landing, Mare Insularum
Apollo XIII 4/11/70	James A. Lovell John L. Sweigert Fred W. Haise	Aborted on the way to Fra Mauro on the Moon due to explosion in Service Module
Apollo XIV 1/31/71	*Alan B. Shepard Stuart Roosa *Edgar D. Mitchell	Third lunar landing, Fra Mauro area
Apollo XV 7/26/71	*David R. Scott Alfred M. Worden *James B. Irwin	Fourth lunar landing, Hadley-Appennines area
Apollo XVI 4/16/72	*John W. Young Thomas K. Mattingly *Charles M. Duke	Fifth lunar landing, Descartes area
Apollo XVII 12/7/72	*Eugene A. Cernan Ronald E. Evans *Harrison H. Schmitt	Sixth lunar landing, Taurus-Littrow Valley

*Indicates an astronaut who landed on the Moon.

Aquaculture

Archaeological Geoscience

See Careers in Oceanography.

artifact tool, artwork, or other object made by humans

Archaeological geoscience is the application of the earth sciences to archaeology, the study of past human cultures. It helps archaeological investigators trace vanished landscapes, establish the dates of the sites they discover, and identify the sources of ancient **artifacts.**

Geology and archaeology both developed in the first half of the 1800s. For many years, the two sciences shared a common goal: the quest for knowledge about ancient peoples. Archaeology borrowed many methods and ideas from geology; geologists and other earth scientists are still helping archaeologists piece together the human past.

Links Between Geology and Archaeology. Geology and archaeology have much in common. Both involve studying the ground and delving below its surface to examine material that has been deposited over time and the rocks and other mineral matter associated with it. Geologists and archaeologists alike interpret strata, or layers, of **sediment** and **sedimentary rock.** Both types of investigators seek to determine the age of the materials they find, although archaeologists examine artifacts, whereas geologists study fossils.

sediment soils, rock particles, and other materials that are deposited over time and make up the ground, whether on dry land or at the bottom of a body of water

sedimentary rock rock formed from deposits of sediments (soils, rock particles, and other materials) over long periods of time

stratigraphy branch of geology that deals with strata, or layers, of rock

geochemical related to the chemical composition of the Earth's materials

geophysical related to the physical characteristics and structure of the Earth

Stratigraphy is an important element in the fields of geology and archaeology. Experts in both fields have learned to work with incomplete records, since sediments and strata in many places are confused by erosion and movements withing the Earth. Like geologists, archaeologists may use drilling, **geochemical** and **geophysical** analyses, and similar techniques to gather data. Geologists and archaeologists alike use maps and cross sections to organize the information they find.

Reconstructing Hidden Landscapes. Archaeologists who want to understand an ancient community need to know what the landscape was like when people lived there. Many of these landscapes, however, have since been

An archaeologist records site data about a core sample of sediment for later analysis.

folding movements of the Earth's crust that cause rocks layers to bend into folds

faulting movements of the Earth's crust that produce a fault (a break or crack in the Earth's rock layers that creates a surface along which the rocks slide)

geomorphology study of the Earth's surface and the forces that shape it

sedimentology study of sediments

tectonic having to do with shaping the Earth's crust

radiometric dating measuring amounts of radioactive material that decays at a known rate in order to establish the age of a sample

Finding Lost Harbors

The location of ancient Carthage—the North African city that waged war with Rome—was never in doubt. Historians could not agree, however, on the number, shape, and location of its harbors. In the early 1990s, investigators used core samples to determine how Carthage's coastline must have looked thousands of years earlier. The study showed the types of harbors that could have existed then and their probable locations. Combining geological and historical information, these researchers outlined just how a famous Roman assault on the city in 146 B.C. might have happened. Their work demonstrates how the earth sciences can fill in some of the gaps in the history of ancient civilizations.

altered by erosion, changes in sea level, **folding** and **faulting**, and other processes that shape the face of the Earth. This is especially true along rivers and around large bodies of water—places where early cultures often thrived.

Archaeologic geoscientists use techniques from **geomorphology** and **sedimentology** to reconstruct these lost landscapes. Geomorphic surveys show how the land, rivers, and sea changed over time. Investigators look for evidence of three processes: worldwide rise or fall of sea levels, **tectonic** movements of the rock layer, and the deposit or erosion of sediments.

Core drilling—obtaining a vertical sediment or rock sample—is one of the most precise techniques for studying ancient coastal landscapes. Scientists often use **radiometric dating** of core material to determine the age of layers in these sediment samples.

In the last few decades of the 1900s, sedimentology became a significant part of archaeological excavation. Soil analysis has provided clues to the ancient natural environment of a site as well as to agriculture and other long-ago human activities there.

Advanced technology has given archaeological geoscience new tools. Remote-sensing instruments in satellites, for example, measure radar, infrared, and visible light radiation. In 1981 a radar imaging system aboard the space shuttle *Columbia* revealed river channels shallowly buried under the desert in Egypt and the Sudan. These images, which could not have been obtained on the ground, led archaeologists to new sites.

Tracking Artifacts to Their Sources. Archaeologic geoscientists use a variety of techniques to "fingerprint" artifacts and identify their sources. Such investigations are called provenance studies (*provenance* means "origin" or "source"). Marble is a good example of a material that can be traced to its source. It occurs in many varieties, each of which has distinctive geologic and chemical characteristics. In some cases, scientists are able to analyze a marble statue from the Mediterranean region, compare their findings with known sources of marble, and pinpoint the quarry that supplied the marble for the statue.

The results of provenance studies can surprise even experts. For example, on the Plain of Thebes in Upper Egypt, researchers examined two larger-than-life stone figures of an ancient pharaoh. The statues proved to be made of quartzite stone from a quarry more than 400 km down the Nile River, although there were closer sources upstream. Archaeological geoscientists have identified the sources of objects made of clay, silver, lead, and copper as well as stone.

Establishing Dates. Archaeologic geoscientists sometimes use volcanic ash to determine the age of a site. If they find a layer of ash in an area, they can use radiometric dating to establish when the eruption occurred and to determine the age of the surrounding material.

Some archaeological sites contain disturbed and broken layers known as "destruction layers." Scientists can examine these layers to see whether an earthquake caused the damage to buildings and other relics. However, not all such damage is due to earthquakes. Floods, fires, and warfare also cause destruction layers.

Archaeologic findings can also contribute to knowledge in the earth sciences. Archaeomagnetic dating is based upon the fact that Earth's magnetic poles are not fixed but move from place to place over time. If an object became magnetized at some point in the distant past and has not moved since then, a scientist can determine where the magnetic poles were located when it was magnetized. If the object's age can be established, another data point is added to the archaeomagnetic dating system.

Geomorphology and sedimentology, provenance studies, and dating techniques are some of the tools from the earth sciences that cast light on ancient cultures. Studies of the Earth and of the human past come together in archaeological geoscience. SEE ALSO CARBON-14 DATING; SEDIMENTOLOGY; STRATIGRAPHY; STRUCTURAL GEOLOGY.

Arthropods

invertebrate animal without a backbone

GT See Geologic Timescale on page viii.

appendage limb or similar body part

larva immature form of an animal that is essentially different from the form of the parents

Arthropods make up a major phylum, or classification, of the animal kingdom. This phylum includes an enormous array of **invertebrates,** such as insects, spiders, scorpions, ticks, millipedes, centipedes, shrimps, crabs, and lobsters. The earliest known arthropod fossils come from early in the Cambrian period^{GT} of the Paleozoic era, more than 540 million years ago.

General Characteristics. The term *arthropod* comes from Greek words meaning "jointed leg," a basic characteristic of this group of animals. In addition to jointed legs, all arthropods have a stiff body covering called chitin, which forms an exoskeleton (external skeleton) around them. The exoskeleton provides protection for soft body parts and support for muscles.

Arthropods share a number of other characteristics. Their bodies are divided into distinct segments. In most arthropods, certain segments combine into functional groups—a head; a thorax, or body; and an abdomen, or tail region. They have symmetrical pairs of legs and other **appendages,** such as antennae, that are used for walking, feeding, swimming, breathing, mating, and other specialized purposes. Arthropods have a complex nervous system with a brain and nerve cord, as well as a complete digestive system.

The stiff exoskeleton protects arthropods, but it cannot expand to accommodate growth. Arthropods do, however, have the ability to molt, or shed their exoskeletons, at intervals throughout their lives. Beneath the old exoskeleton is a new one that is larger and softer. As the arthropod grows to fill it, this new exoskeleton hardens to form a protective covering once again. Many arthropods, such as butterflies, also undergo a metamorphosis, or complete change in form, from a **larva** to an adult.

A fossil shrimp, *Aeger tipularis,* from the Jurassic period.

A Cambrian period trilobite fossil, *Moducia typicalis,* measures 35 mm. It was found in Utah.

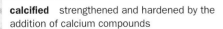

calcified strengthened and hardened by the addition of calcium compounds

Early Arthropods.   A variety of arthropods lived in Cambrian seas more than 540 million years ago. Although there are fossils of several groups of arthropods, most Cambrian marine deposits are richest in crustaceans and an extinct group known as trilobites. These two groups are well preserved because their exoskeletons were **calcified**; there was a great diversity of species (250,000 to 2,000,000 for trilobites alone); and their populations were very large. Crustaceans of the period included ostracodes, tiny arthropods with clamlike exoskeletons. Trilobites had long oval skeletons with many segments.

Arthropods were the earliest examples of many evolutionary developments. They were the first animals to develop acute eyesight. By early Cambrian

compound eye eye made up of many tiny, independent visual units

times, many arthropods had **compound eyes,** as do modern insects. Arthropods were the first animals to make the transition from the oceans to land. Some crustaceans, chelicerates (the group that came to include spiders and scorpions), and myriapods (now including centipedes and millipedes) had managed this feat by the Silurian period, 438 million years ago. Once established on land, insects developed wings and became the first animals to fly. By the end of the Paleozoic era, some 245 million years ago, insects had become the most numerous animals on land, a position they still hold.

Although many arthropod groups became extinct at the end of the Paleozoic era, remaining groups continued to evolve. Modern types of crabs, shrimps, and lobsters appeared by the Jurassic period, 208 million years ago. Within another 50 to 60 million years, around the middle of the Mesozoic era, the modern groupings of arthropods were taking shape.

Evolutionary Success. Modern arthropods live in a wide variety of environments—from the deepest regions of the ocean to shallow water, in freshwater such as lakes and rivers, and in all land environments. Some arthropods have gills, like those of fish, that let them breathe in water. Others breathe directly through air tubes or by means of lungs. Arthropods move by swimming, walking, crawling, and flying. They may be herbivores—feeding only on plants—or carnivores—feeding on the blood or flesh of other animals. Some arthropods are parasites, living directly on or inside the body of another animal.

Many arthropods are a major source of food for other animals. Frogs, bats, and certain other animals feed on various types of insects. In certain cultures, even people eat insects. Crustaceans are a popular food in many cuisines, so shrimps, crabs, lobsters, and crayfish are harvested commercially in numerous coastal areas around the world.

Arthropods can be both helpful and harmful to humans. Some insects eat plants and can destroy farm crops. Other arthropods, such as certain species of ticks and mosquitoes, can spread disease. Some spiders are poisonous, and bees and certain types of ants can inflict painful and sometimes lethal stings on humans. Many other arthropods are beneficial. Several crustaceans and some other groups are a prized source of food, while others feed on harmful arthropods, pollinate plants, or produce useful substances such as honey and silk. SEE ALSO INVERTEBRATES.

Asteroids

Asteroids are objects that revolve around the Sun, as planets do, but are much smaller than planets. The largest known asteroid, Ceres, is nearly 1,000 km in diameter; others are the size of boulders or even smaller. Most asteroids endlessly circle the Sun between the orbits of Mars and Jupiter. Some, however, approach or cross Earth's orbit. Scientists study the role that asteroid collisions have played in shaping the solar system—including, perhaps, the history of life on Earth.

astronomical unit (AU) Earth's distance from the Sun, or approximately 150 million km

Asteroids and Their Origin. Astronomers discovered Ceres, the first known asteroid, in 1801. They expected to find a planet between Mars and Jupiter, since planets tend to be regularly spaced, but instead they located a number of asteroids in a belt that extends from 2.2 to 3.2 **astronomical units** from the Sun. Known as the main belt, this area accounts for most asteroids. Other asteroids, called the Trojans, revolve around the Sun at the same distance as Jupiter. A small percentage of asteroids have orbits that carry them into the inner solar system, where they may collide with Earth.

Close-up of asteroid 243 Ida and its moon.

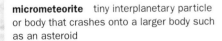

micrometeorite tiny interplanetary particle or body that crashes onto a larger body such as an asteroid

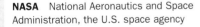

NASA National Aeronautics and Space Administration, the U.S. space agency

Asteroids are leftovers from the formation of the solar system. The planets formed when smaller bodies called planetesimals collided with each other and were held together by gravity. Not all planetesimals became part of planets, however. Those that remained among the planets are asteroids. Others were located in the extremely cold outer fringes of the solar system. These contain ice, and when orbital changes carry them toward the Sun, trailing tails of vapor, they are called comets.

Getting a Closer Look. In 1991 astronomers got their first close look at an asteroid when the *Galileo* spacecraft, on its way to Jupiter, flew past the asteroid 951 Gaspra. The asteroid is covered with small craters created by impacts with interplanetary rubble over millions of years. Slight color differences on its surface made scientists think that its soil had been "space-weathered" by **micrometeorites** and radiation.

Two years later, images of the large, heavily cratered asteroid 243 Ida revealed surprising new information. A small, egg-shaped satellite revolves around the asteroid. Scientists dubbed this moonlet Dactyl. By studying its orbit around Ida, scientists determined that Ida contains little metal and, most likely, a large amount of empty space. The asteroid may really be a pile of rubble around two large pieces of rock.

In early 1996, **NASA** launched the Near Earth Asteroid Rendezvous (NEAR) spacecraft. The spacecraft flew past asteroid 253 Mathilde in 1997, producing scores of images of the asteroid. NEAR is scheduled to reach the near-Earth asteroid 433 Eros in 1999. The spacecraft will orbit Eros for a year, relaying detailed information about its size, shape, structure, mass, magnetic field, and chemical makeup to scientists on Earth. The close-up exploration of asteroids will continue with Deep Space 1, a spacecraft launched late in 1998, that will fly close to the near-Earth asteroid 1992 KD in July 1999. If NASA extends the mission, Deep Space 1 will fly close to two asteroids in 2001—Wilson-Harrington in January and Borrelly in September.

Asteroids and Meteorites. Going into space is one way to study asteroids. Another is to examine pieces of asteroids that have landed on Earth. Chunks of material from space that fall toward Earth are called meteoroids. They form when asteroids in the main asteroid belt collide. Such collisions produce fragments whose disturbed, irregular orbits eventually cross the orbit of Earth. Many meteoroids burn up in Earth's atmosphere; these are called meteors. Occasionally, however, something survives to land on Earth and is known as a meteorite.

Meteorites fall into three broad categories: stones, irons, and stony-irons. Meteorites in two of these categories present an interesting puzzle. Stones are the most common of the meteorites, and most stones are of a type known as chondrites. Most asteroids appear chemically similar to the carbonaceous chondrites, which are thought to be the most common rock in space. None of the main-belt asteroids, however, seem chemically similar to the more common "ordinary" chondrites. Furthermore, many main-belt asteroids appear to be chemically similar to the much rarer stony-iron meteorites.

The *Galileo* flyby of Gaspra, with its evidence of space-weathering on the asteroid's surface, may have helped unravel the mystery. It seems that the space weathering process may alter the apparent chemical nature of the surface of chondrite material so that it resembles that of stony-iron material.

The Spaceguard Survey

The search for objects that might strike Earth is called the Spaceguard Survey. Agencies such as the U.S. Air Force, the Jet Propulsion Laboratory (sponsored by NASA and run by the California Institute of Technology), and the Lunar and Planetary Laboratory at the University of Arizona are working together on the survey. By 2007 they hope to have identified 90 percent of all near-Earth asteroids and other objects that measure at least 1 km in diameter. By mid-1998 the group had found 29 such objects but estimated that between 1500 and 2000 of them exist. The Spaceguard Survey will not only help scientists choose targets for future space exploration missions but may also help protect the Earth from an asteroid impact.

Asteroids and the Earth. Scientists believe that nearly 2,000 objects larger than 1 km in diameter are orbiting the Sun on paths that could bring them close to Earth. These NEOs, or near-Earth objects, include both asteroids that were knocked out of the main belt by collisions and "dead" comets that have lost their ice and no longer show tails. Live comets also occasionally approach the Earth. Earth's atmosphere provides protection from objects up to about 100 m in diameter. Larger NEOs, however, could actually strike the planet.

Such impacts have occurred before. Craters on the Moon and on other bodies in the solar system are evidence of much bombardment from space. Scientists have concluded that around 4 billion years ago, when the Moon received an especially heavy bombardment, many large NEOs crashed into the Earth as well. Few large craters remain because erosion, ocean formation, and other changes in the planet's surface have erased them. Still, scientists who have studied existing impact craters believe that the Earth is hit by a 10-km NEO every 100 million years or so.

One lively area of research involves the effect of bombardment on planetary history. Life probably could not arise until after the heavy bombardment that accompanied the birth of the solar system had ended. Since the 1980s, some geologists have suggested that huge NEO impacts—and the global tidal waves and dust clouds they created—were responsible for large-scale extinctions, such as the disappearance of the dinosaurs.

Every several hundred thousand years, the Earth is struck by a NEO in the 2 km range. In addition to causing considerable destruction in its immediate area, such an impact could produce worldwide climate-changing effects. Several popular books and movies in the 1990s dealt with the possibility of NEO impacts, and although these treatments were fictional, scientists and the military have begun to discuss ways of predicting and preventing future impacts. SEE ALSO COMETS; EXTINCTIONS; IMPACT CRATERING; METEORITES; SOLAR SYSTEM.

Asthenosphere

plate tectonics theory used to explain continental drift (the movement of continents over the surface of the Earth) and other geological processes

One component of the Earth's interior, the asthenosphere is part of the upper mantle, a region below the Earth's crust. The asthenosphere lies on average about 75 km beneath the surface of the Earth.

Composed of partially molten rock, the asthenosphere plays an important role in **plate tectonics.** Above the asthenosphere is the lithosphere, a rigid layer of mantle and crust broken into enormous plates. These plates are able to move over the more plastic, or pliable, asthenosphere. Lithospheric plates on the ocean floors are denser than the hotter material in the asthenosphere. Very slowly, these plates sink back into the asthenosphere along deep-sea trenches. SEE ALSO EARTH, STRUCTURE AND COMPOSITION OF; MANTLE; PLATE TECTONICS.

Astrobiology

evolution changes in groups of related organisms occurring over time

organic related to carbon compounds, which are found in living things; living or formerly living

nucleic acids class of complex chemicals found in the cells of living organisms and important in determining how those organisms develop and function

Astrobiology is the study of the origin, **evolution,** and future of life on Earth and elsewhere. Among the fundamental puzzles astrobiologists consider are how life began and evolved on Earth and whether there is life elsewhere in the universe.

Several approaches are used to study the origin of life on Earth. Astrobiologists examine the geologic record to see what the environment of the earliest life-forms on Earth may have been like; then they try to recreate such an environment in laboratories. Studies of this sort have shown how **organic** chemicals important to life, such as proteins and **nucleic acids,** may have arisen and eventually led to living organisms. Another approach compares the genetics of different organisms. Genetic similarities and differences can show how organisms are related to each other and how their characteristics were inherited from their earliest ancestors. These genetic studies appear to suggest that all organisms on Earth are descended from heat-loving, hydrogen-eating microorganisms that existed 3.5 billion years ago.

Because life developed very quickly after the Earth formed, scientists suspect that life could be common in the universe. Scientists have determined that there are planetary systems around other stars, but not whether there are planets similar to Earth. To determine which planets may be able to support life, and what forms such life might take, astrobiologists study patterns of evolution. The fossil record reveals major evolutionary steps for life on Earth, such as the development of multicellular organisms and the spread of life into widely varied environments across the globe. Astrobiologists use such information to focus their search for signs of past or present life elsewhere than on Earth, such as on the planet Mars and on Europa, one of Jupiter's moons. SEE ALSO EXTRATERRESTRIAL LIFE, SEARCH FOR; FOSSILIZATION AND THE FOSSIL RECORD; GALILEAN SATELLITES; LIFE, EVOLUTION OF; LIFE, ORIGIN OF; MARS; PLANETARY SYSTEMS, OTHER.

Astronauts

See Apollo Astronauts.

Astronomy, Observational

Astronomy is the study of the universe and the stars, planets, and other heavenly objects. For thousands of years, people have gazed up at the sky, observing the positions and movements of the Sun, Moon, and stars. Such observations led to the development of the science of astronomy. Today, professional and amateur astronomers, as well as casual backyard observers and hobbyists, all enjoy watching interesting objects in the night sky.

Observing with the Naked Eye. Using just the naked, or unaided, eye, most people can see the Moon and the bright planets and stars. Even distant

Time exposure image of the Moon and stars tracing arcs in the sky above part of the Keck Telescopes facilities on Mauna Kea in Hawaii.

and faint stars and constellations, or patterns of stars, are visible in rural and some suburban areas, especially in high desert country. Ancient peoples, looking at these star patterns, visualized the outlines of lions, horses, bears, hunters, and other figures. Many constellations in the sky are named after these figures.

Looking at the stars, one notices that some are brighter or fainter than others. Differences in brightness, or magnitude, depend on the temperature and size of a star and its distance from Earth. Scientists today use a system of magnitudes first devised by the ancient Greek astronomer Hipparchus, who ranked stars according to their brightness.

Sometimes it is possible to catch a glimpse of a "shooting star" in the night sky. A shooting star, or meteor, is a small particle of matter from space that burns and glows as it hurtles through the Earth's atmosphere. Spectacular "meteor showers" occur at certain times of the year. These swarms of often thousands of meteors are the remains of comets—balls of mostly ice, gases, and dust that travel around the Sun in long elliptical, or oval-shaped, orbits.

On rare occasions, comets are visible as they pass by the Earth on their orbits around the Sun. Since the mid-1980s, several comets have received much attention. The most famous of these is Halley's comet, which passes by the Earth about every 76 years.

Some planets, including Venus, Mars, Jupiter, and Saturn, can be seen with the naked eye. Ancient peoples recognized these planets, although they did not know what planets were, because their movements differed from those of the stars. The planets trace distinct paths among the stars, and sometimes they appear to slow down and reverse direction, a phenomenon known as retrograde motion. These apparent patterns actually result from the motion of the Earth around the Sun in relation to the orbits of the other planets.

The largest and brightest object in the night sky is the Moon, the closest natural object in space to Earth. The only objects closer (besides occasional meteors) are artificial satellites, which can often be seen traveling across the night sky. It is very easy to see the different phases of the Moon as it changes each month. With a "full Moon," it is also possible to see dark and light areas on the Moon's surface.

On a clear night, a band of milky light can sometimes be seen stretching across the sky. This is the Milky Way, a galaxy composed of millions of stars and nebulae, or clouds of gas and dust. Our solar system—which includes the Sun, Earth, other planets, and moons—is a tiny part of this enormous galaxy.

Observing with Telescopes and Other Devices. In the 1600s, scientists began to build devices called telescopes—instruments that make distant objects appear nearer, brighter, and more distinct. Binoculars are essentially two low-powered telescopes that are joined side by side to allow the viewer to observe an enlarged image with both eyes.

Various types of telescopes are in use today. The most familiar are optical telescopes, which gather light from distant objects and produce an enlarged image. There are two basic kinds of optical telescopes. Refracting telescopes are tube-shaped devices with a lens at each end. The lens at the end facing the sky gathers light, focuses it, and projects it to the other end of the telescope, where the lens of the eyepiece magnifies the image. Reflecting telescopes have a curved mirror at the back of the tube that gathers and reflects the light to a smaller mirror and then to an eyepiece.

Telescopes allow people to see distant galaxies, nebulae, clusters of stars, and other objects that are not visible to the naked eye. Through a telescope, most of these objects appear as hazes of light and fuzzy spots. However, some, such as the Andromeda galaxy, are bright enough to be visible with ordinary binoculars.

Telescopes also allow people to see closer objects, such as the Moon and planets, more clearly and to distinguish some of the features of their surfaces or upper atmospheres. For instance, through a telescope, the largest explosions ever viewed on another planet—the collisions of Comet Shoemaker-Levy with Jupiter—were observed in 1994. Jupiter is known for its brownish equatorial belts and Great Red Spot, both imperceptible to the naked eye. The famous rings of Saturn and dark and bright markings on Mars, such as Syrtis Major and Solis Lacus, can also be viewed through a telescope.

Atmospheric Extinction

When stars rise and set, their light shines through a thicker part of the atmosphere than when they are high in the sky. This change causes their light to appear to fade, an effect known as atmospheric extinction. Similarly, the Sun looks red at sunset because its blue and green rays are scattered as they pass through this thicker part of the atmosphere, leaving only red light.

A reconstruction of the first telescope used for astronomical purposes. The original was created by Italian scientist Galileo Galilei in 1610. The Cathedral of Florence is in the background.

Observational Limitations.   Several factors limit astronomical observations on Earth. Certain weather conditions, such as heavy clouds, limit our ability to look into space with optical telescopes. In urban areas, observation is also limited by artificial lighting, which produces a lot of glare at night. In addition, the atmosphere acts as a filter, blocking or absorbing certain amounts of light, X rays, and other types of radiation. Despite these limitations, in clear weather, it is almost always possible to enjoy a fascinating view of heavenly objects in the night sky. SEE ALSO COMETS; GALAXIES; GALILEAN SATELLITES; JUPITER AND SATURN; MARS; MERCURY; METEORITES; MOON; PLANETARY RINGS; PLUTO AND CHARON; SATELLITES, ARTIFICIAL; SOLAR SYSTEM; STARS; TELESCOPES; URANUS AND NEPTUNE; VENUS.

Astrophysics

See Cosmology.

Atmosphere, Earth's

The Earth's atmosphere is made up of several layers of gas that are held to the planet by gravity. This envelope of gas supports life, carries wind and weather, and shields the planet's surface and its inhabitants from harmful space radiation.

Structure of the Atmosphere.   The atmosphere is a thin layer spread around a large sphere. Although the atmosphere extends thousands of miles up from the planet's surface, most of the air it contains is near the ground. Its upper levels are extremely thin, fading into the near vacuum of space.

The atmosphere is divided into layers. Closest to Earth is the troposphere. A boundary called the tropopause divides it from the next layer, the stratosphere. Between the stratosphere and the next layer, the mesosphere, is a boundary called the stratopause. A third boundary, the mesopause, separates the mesosphere from the outermost layer of atmosphere, which is called the thermosphere.

The height of the boundaries—and therefore the thickness of the layers—varies by season and by location over the Earth's surface. The average heights

Vertical temperature distribution of Earth's atmosphere.

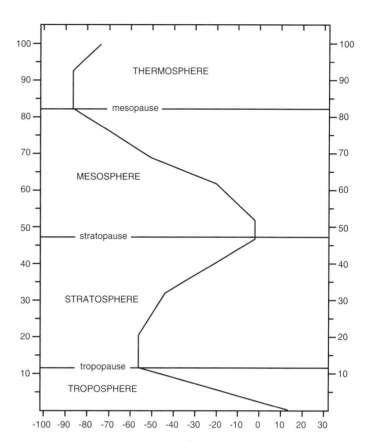

are about 11 km to the tropopause, 47 km to the stratopause, and 82 km to the mesopause. Most of the atmosphere is significantly colder than the Earth's surface temperatures, but the upper stratosphere and lower mesosphere are warmer than the rest of the atmosphere because they contain ozone, a form of oxygen that absorbs ultraviolet radiation from the Sun.

Air pressure is a measurement of the weight of the air above a given point. An instrument called a barometer measures air pressure, using units called millibars (mb); the average air pressure at sea level is 1,000 mb (or 1 bar). Air pressure falls dramatically with height above sea level, dropping to 500 mb at 5 km, and is insignificant by about 30 km.

Chemical Composition of the Atmosphere.   Air is a mixture of gases, primarily nitrogen (just over 78 percent), oxygen (nearly 21 percent), and argon (nearly 1 percent). These proportions are stable, although amounts of other gases are present in varying degrees. Water vapor and carbon dioxide levels affect the climate. Unpolluted air also contains tiny amounts of neon, helium, methane, nitrous oxide, ozone, hydrogen, nitric oxide, ammonia, sulfur dioxide, and other gases. Polluted air contains increased amounts of many of these gases as well as more than 1,000 chemical compounds produced by industry and other human activities.

Most atmospheric gases have no color, odor, or taste. People cannot sense changes in their concentrations. Water vapor is an exception: People are aware of high humidity. People can also sense some of the effects of certain air pollutants.

Studies of the chemical composition of the atmosphere have shown that the proportion of carbon dioxide to other gases is steadily increasing. Human activities—chiefly clearing forests and burning fossil fuels—have caused this increase. By analyzing ancient air that was trapped in bubbles in polar glaciers, scientists determined that between 1750 and 1990, atmospheric carbon dioxide increased by 25 percent, and the annual increase was getting larger each year. Because carbon dioxide absorbs some of the radiation emitted by the ground and returns heat energy to the ground, it affects the energy balance on Earth. We are not yet sure what effect increasing carbon dioxide levels will have on the global climate, but it is certain to cause some change.

How the Atmosphere Formed.   Using air trapped in bubbles of Antarctic ice, researchers have studied the makeup of Earth's atmosphere over the past 170,000 years. For earlier periods, they must draw their conclusions from knowledge about the properties of Earth materials and the formation of the Earth.

The Earth is not made out of the same materials as the Sun, which is mostly hydrogen. Like the other inner planets, the Earth was formed from a collection of rocky, solid little planets without atmospheres, at a time when most of the gases in the solar system had been swept into the Sun or into the outer planets. The Earth's atmosphere developed during the bombardment of the growing Earth by little planets. The collisions produced heat, stimulating chemical reactions that drove gases and liquids out of the rock to the planet's surface, where gravity held them in place. At first the atmosphere was hot and heavy, consisting mostly of steam and carbon dioxide.

When the bombardment ended and surface temperatures fell, water vapor condensed out of the atmosphere to form the oceans. Over hundreds of millions of years, carbon dioxide from the atmosphere reacted with seawater and surface materials to form solid carbonate minerals. This process left nitrogen as the most abundant atmospheric gas. All of these changes took place before the formation of the oldest rocks that have yet been found, so there is no direct evidence that they took place.

However, in rocks about two billion years old, scientists have found traces of chemical reactions involving oxygen, which can only have been produced by green plants through the process known as photosynthesis. The increase in oxygen as a result of biological processes led, over millions of years, to the formation of the atmosphere in which we live. This interaction between lifeforms and the planet is the most significant event in the geological history of the atmosphere. SEE ALSO CLIMATE CHANGES; CLIMATOLOGY.

Atmospheres, Planetary

mass   amount of matter that causes an object to have weight when it is in an area influenced by gravity; commonly measured in kilograms

A gaseous envelope, or atmosphere, surrounds each of the nine planets in our solar system, at least four of their moons, and most comets. All of these atmospheres are made mainly from the same six elements—hydrogen, helium, oxygen, carbon, nitrogen, and sulfur—but they differ in the amounts of each of these elements they contain. They also differ in **mass,** temperature, and the degree to which they interact with the planets they surround.

The atmospheres of the largest gaseous planets merge smoothly with interiors made of the same mix of gases; pressures inside these "gas giants" reach millions of times the atmospheric pressure at sea level on Earth. By contrast, the atmospheres of the Moon and Mercury are little more than individual molecules hopping around the surface (and being lost in space) at densities (pressures) too low to collide with one another. On other planets, including

Earth, the atmosphere rests on solid or liquid surfaces and interacts chemically with the planetary surface.

Hydrogen and helium are the most abundant elements in the atmospheres of gas giants such as Jupiter and Saturn, while the compound carbon dioxide (CO_2) is the most abundant ingredient in the atmospheres of Venus and Mars. Nitrogen is the most common element in Earth's atmosphere, followed by oxygen. However, without water in liquid form, Earth's atmosphere would have 40 to 100 times the mass of its actual atmosphere, in CO_2 alone. Because CO_2 dissolves readily in water, CO_2 from Earth's atmosphere has reacted with calcium-rich rocks to form calcium carbonate (limestone).

Weather and Climate. Atmospheres move constantly, creating and changing weather and climate, because they are heated unevenly. Sunlight is the main heat source for most planetary atmospheres; it heats only the dayside of a planet and heats the equator more than the poles. **Convection** carries heat upward from a planet's surface and toward the poles, reaching altitudes where the energy escapes into space. Thunderstorms and hurricanes are examples of small- and intermediate-scale forms of convection that transport heat vertically. Large-scale convection transports heat laterally, or sideways. An example is Earth's Hadley circulation, which carries heat from the equator to about 30° latitude. From there, large rotating masses of air called cyclones and anticyclones carry it to the poles.

These same weather elements exist in other planetary atmospheres, but wind speeds are generally larger and storms last much longer than on Earth, even on the outer planets, where sunlight is much weaker. Farther from the Sun, there is less small-scale convection, a condition that may allow the need for large-scale convection to build up. A dramatic example is Jupiter's Great Red Spot, a huge storm that was discovered shortly after the invention of the telescope 300 years ago and is still active today. SEE ALSO ATMOSPHERE, EARTH'S; COMETS; HURRICANES; SOLAR SYSTEM.

Atomic Energy

See Nuclear Energy.

Auroras

convection circulatory movement in an unevenly heated fluid (liquid or gas); cooler material is denser and sinks in an area influenced by gravity, while warmer material is less dense and rises

Auroras are displays of light that appear in the night sky at high latitudes. Found in the northern and southern hemispheres, they generally occur in narrow belts around the polar regions. Auroras in the northern hemisphere are called the Aurora Borealis, or "northern lights." Those in the southern hemisphere are called the Aurora Australis, or "southern lights."

Auroras take varying forms, including arcs, streamers, and bands of light that resemble hanging curtains. They appear in shades of green, red, yellow, blue, and violet. The colors in auroras depend on the types of gases in the atmosphere where they occur.

Auroras are produced by fast-moving, electrically charged particles from the Sun. As these particles sweep past Earth, they are drawn into the upper atmosphere at the polar regions by the planet's magnetic field. These particles strike other particles in the atmosphere, causing a discharge of electricity and the generation of variously colored light. Although auroras occur continually near the polar regions, they appear less frequently in regions farther from the poles.

The light-producing electrical discharges that cause auroras generally occur about 250 km above the Earth. They continue to grow brighter until they reach about 115 km above the Earth, at which point the light disappears. On rare occasions, auroras may extend thousands of kilometers high.

Auroras are one of the most beautiful natural phenomena on Earth. In recent years, scientists have discovered that auroras also occur on other planets in our solar system that have both an atmosphere and a magnetic field, such as Jupiter, Saturn, Uranus, and Neptune. SEE ALSO ATMOSPHERE, EARTH'S; MAGNETIC FIELD, EARTH'S; PLANETOLOGY, COMPARATIVE; SUN.

Bagnold, Ralph Alger
1896–1990
British engineer and scientist

geomorphology study of the Earth's surface and the forces that shape it

During Ralph Alger Bagnold's outstanding military career, he also made major contributions to **geomorphology** and engineering. Bagnold's love of the desert led him to examine closely the movement of sand by wind, and he also investigated the properties of soil and other solids carried in water. During and after World War II (1939–1945), Bagnold put his scientific knowledge to practical use by perfecting techniques of desert travel and solving problems connected with the physical properties of blown sand.

Into the Army. Born in Stoke, Devonport, England, Bagnold was always interested in scientific exploration. He wrote in his autobiography: "My main urge from boyhood onward, was curiosity [along with] a vague longing to discover something new."

In 1914 Bagnold entered the army as a military engineer. He saw active service in World War I (1914–1918) and then earned an engineering degree from Cambridge University while on educational leave. Upon his return to active military service, Bagnold was posted to Ireland, Egypt, India, and Hong Kong before ill health forced him to retire in 1935.

The Lure of the Desert. While in Egypt, Bagnold made a number of trips into the North African desert. There he developed techniques and equipment, including the sun compass and the closed-vehicle cooling system, to improve desert travel. His 1935 book, *Libyan Sands: Travel in a Dead World,* documented these expeditions.

During World War II, Bagnold—his health restored—was called into service in the British army in East Africa. His experience in desert travel qualified him to organize and lead the Long Range Desert Group, which operated behind enemy lines with great success. After the war, Bagnold continued to follow his interest in the physical properties of windblown sand, designing experiments and instruments to help him in his studies. His 1941 book, *The Physics of Blown Sand and Desert Dunes,* was the first detailed treatment of the subject and is still a classic in geomorphology.

During the 1950s, 1960s, and 1970s, Bagnold published a number of scientific papers on the properties of solids carried by water. He also traveled widely to advise project engineers who were working with blown sand. Among the many scientific and engineering honors he received were the Gold Medal of the Royal Geographical Society, membership in Great Britain's Royal Society, and the Penrose Medal of the Geological Society of America.

Big Bang Theory

See Cosmology.

Biomass Fuels

Biomass, in the form of living and nonliving plant matter and animal wastes, is the most abundant and accessible renewable source of fuel in the world. It includes the stems, branches, and roots of plants; wood products such as sawdust and paper; food processing wastes; and manure.

Plants use energy from the Sun to convert water and carbon dioxide gas from the atmosphere into chemical compounds that enable the plants to grow and build more biomass. This process is called photosynthesis. The chemical compounds stored in the plants are a form of energy that is released when the plant matter is burned.

Biomass fuels were the only source of energy in human societies for thousands of years. In the mid-1800s, **fossil fuels** began to replace biomass fuels as the primary energy source. Today, because of increasing concern over pollution and the diminishing long-term supply of fossil fuels, biomass energy is being rediscovered.

Actual use of biomass fuels for energy varies from country to country. In the United States, only about 3 percent of total energy production comes from biomass fuels. In some countries, however, the figure is close to 40 percent. The most common biomass resources currently used for energy are wood harvested from forests and agricultural residues. Other biomass resources include **peat,** sawdust, newspapers, tree trimmings, and livestock wastes.

Biomass wastes and residues can supply only a limited portion of the world's total energy demand. Moreover, the cutting of forests and excessive removal of natural plant growth may contribute to climate change and have other harmful environmental effects. One solution is to grow crops of trees and grasses for use as biomass fuel. Scientific research is currently making great strides in increasing the potential for the environmentally sound production of large amounts of biomass.

The most frequent use of biomass fuels today is in the production of heat and steam. The biomass is burned in devices from simple wood stoves to huge boilers capable of generating enormous amounts of steam and heat. Some of the heat and steam produced is used to generate electricity. Biomass is also used to produce liquid fuels, such as ethanol, for automobiles and other vehicles.

Biomass fuels are readily available to any country with biomass wastes and land to grow biomass crops. The development of new technologies to convert biomass fuels into liquid fuels and electricity will make them increasingly important in the future as a replacement for fossil fuels. SEE ALSO CLIMATE CHANGES; ENERGY USE AROUND THE WORLD; FOSSIL FUELS; POLLUTION OF THE ATMOSPHERE.

fossil fuel substance such as coal, oil, or natural gas, found underground in deposits formed from the remains of organisms that lived millions of years ago

peat dark brown or black residue of plant material that has partially decomposed (rotted) underwater

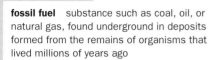

Biosphere

ecosystem system formed by the interaction of a group of organisms and their physical environment

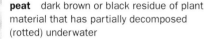

Earth's life-supporting envelope, or biosphere, is the land, water, and air in which life exists. This world of living things includes all of the Earth's **ecosystems** and is the largest unit in the organization of life.

The biosphere is constantly changing. Sudden natural disasters such as wildfires, floods, or volcanic eruptions can destroy local portions of the biosphere. Other changes occur more slowly, but their effects may be great. For example, the release of chemical pollutants into a river may seem to have little immediate effect; in the long run, however, the pollutants may damage many communities of living things. Because human activities are speeding up the rates of change in many parts of the biosphere, it is important to understand how the biosphere works, to monitor it, and to find ways of predicting how present trends and activities will affect it in the future.

Physical Structure of the Biosphere

The biosphere is divided into three main layers. These are the hydrosphere (water), the atmosphere (air), and the lithosphere (earth).

The Hydrosphere. The hydrosphere includes all of the world's water: oceans, rivers, lakes, groundwater, snow, ice, and glaciers. The biosphere penetrates them all—that is, there is life in all of these areas. Life is most diverse and widespread in the oceans. It is found from the surface layers that cover 71 percent of the Earth to the deepest trenches 10,000 m below the surface and from the icy polar waters to the warm tropical seas.

The greatest concentration of life in the hydrosphere consists of organisms that float or swim through the water, from microscopic **plankton** to giant whales. Other organisms live on the ocean bottom, the benthos, where water and land meet. These plants and animals may be supported by rock or earth but carry out vital functions, such as eating and breeding, in the water. Still others—tiny, specially adapted organisms called neuston—live on or in the sea's surface, where water and air meet.

plankton usually minute plant and animal life that floats or swims at or near the surface of water

The Atmosphere. The atmosphere is an envelope of gas that reaches from the surface of the Earth to about 30,000 km above it. Life is common only in the innermost layer of the atmosphere, the troposphere, where air currents are active and significant amounts of moisture are present.

It takes energy for anything to become airborne and stay aloft. Organisms that spend time in the atmosphere must either work at flying or be small enough to float on air movements. Insects, birds, and bats have developed powered flight. Life in the atmosphere, however, consists mostly of tiny things such as aerial plankton or plant seeds, spores, and pollen grains. For them the atmosphere is not so much a habitat as a means of spreading, and coming down may be more difficult than rising.

No organism spends its whole life cycle entirely within the atmosphere. Many function in the atmosphere but are supported by the lithosphere. Plants, for example, receive not only support but also water and nutrients from the earth while they exchange gases in the atmosphere. Through them matter and energy move across the lithosphere-atmosphere boundary.

mantle region of the Earth between the molten core and the outer crust

sediment soils, rock particles, and other materials that are deposited over time and make up the ground, whether on dry land or at the bottom of a body of water

The Lithosphere. The biosphere does not penetrate very deeply into the lithosphere, which is composed of the Earth's crust and uppermost **mantle**—although traces of life can be found quite deep within the crust. Life flourishes, however, in the zones of transition where the lithosphere meets the hydrosphere and atmosphere. When the lithosphere is covered with water, the transition zone often consists of **sediment** into which organisms can burrow. On land it may consist of broken rock or of soil.

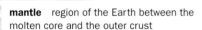

Cycles and Interactions

The biosphere is a dynamic zone of movement and interaction. It plays an essential role in the circulation of materials across the boundaries that separate earth, water, and air. The existence of individual organisms and the ecosystems in which they live depends upon a number of cycles, such as the water cycle, in which water vapor rises into the atmosphere to fall as rain that refills the lakes and seas that produce the vapor. Other cycles move carbon, nitrogen, phosphorus, and various minerals through the biosphere.

Effects produced by the biosphere sometimes reach beyond its own physical limits. For example, massive amounts of carbon and calcium, once removed from seawater to construct the skeletons of ancient organisms, now lie deep within the Earth's crust in the form of limestone. Auroras in the upper atmosphere are an interaction between the Earth and the Sun that depends upon oxygen produced by living plants far below, in the biosphere.

Ecological Organization of Life

A major goal of ecologists is to understand the organization of the biosphere. They have learned that life in the biosphere is hierarchical, or organized into a series of levels. Each level includes many organisms that also belong to the level below it. Individual organisms, for example, form groups called species populations. Various species populations interact to form communities that are linked by food chains or webs—for example, spiders eat insects, then birds eat spiders, and insects eventually eat dead birds.

Ecologists group these communities into larger ecological units called provinces. A province is an area defined by geography and climate, with barriers that prevent its particular pattern of biological diversity from spreading. Although it is not easy to recognize ecological units larger than provinces, ecologists sometimes use terms such as realm, region, division, and domain when classifying or mapping large-scale biological diversity.

Diversity in the Biosphere

Beyond the basic question of how life first arose, the most pressing scientific question about the biosphere concerns the great diversity of life—in other words, why there are so many kinds of plants and animals. Scientists know that there are many more species than they have been able to identify.

The alarming decline in this diversity through human activities raises other questions, such as how diversity can be protected and what is the best way to measure it. Currently scientists measure diversity by listing species, but other methods may be more helpful in both monitoring and preserving diversity. Instead of counting species, scientists may focus on diversity of physical structure, genetics, or function. However it is measured, though, diversity is the aspect of the biosphere about which we understand the least.

History of the Biosphere

The ability to change is a basic property of the biosphere. The biosphere has evolved over time along with the physical Earth. Each has affected the other. Life evolved, to meet the physical conditions of the geosphere, but as it evolved, it also created changes in the hydrosphere, the atmosphere, and the lithosphere.

Life probably began in the hydrosphere. Fossil traces of single-celled marine microorganisms similar to bacteria have been discovered from about 3.5 billion years ago. By about 570 million years ago, living microbes had appeared on land.

At the time of life's origin, the atmosphere was composed mostly of carbon dioxide. The shift to an oxygen atmosphere was one of the most important events in the history of life and the Earth. It opened up a range of new possibilities for living organisms, which became more complex. As single-celled

Living on the Fringe

Scientists are especially interested in the plants and animals that live in extreme conditions on the fringes of the biosphere. For example, in places on the deep ocean floor where the temperature would otherwise be close to freezing, hot, chemical-laden water gushes out of vents. Whole communities of creatures have adapted to live in this water. Unlike most organisms on Earth, whose energy ultimately comes from the Sun through photosynthesis, these vent organisms operate on chemical energy by processing sulfur from the water. Scientists study these and other "fringe" organisms to learn about the limits of the biosphere.

life-forms had done, new multicelled organisms diversified and formed communities in the sea before emerging onto land.

The history of the biosphere is one of constant minor change with occasional large-scale reorganizations. The reorganizations follow mass extinction events, when many kinds of organisms suddenly die out. After such events, new species evolve and new communities form, repopulating the biosphere. SEE ALSO ATMOSPHERE, EARTH'S; EXTINCTIONS; HIGHER LIFE-FORMS, EARLIEST EVIDENCE OF; HYDROSPHERE; LIFE, ORIGIN OF; LITHOSPHERE; SOILS, FORMATION OF AND TYPES.

Birds

extinct having died out completely, leaving no direct descendants (refers to a specific grouping of organisms)

paleontologist scientist who studies prehistoric plants and animals through fossils and other remains

Birds are the only feathered animals. Most modern categories of birds evolved between 65 and 35 million years ago, although the first birds—now **extinct**—lived long before that. Those birds may be linked to extinct reptiles, especially certain dinosaurs. **Paleontologists** are debating whether birds evolved from dinosaurs or whether both evolved from some common ancestor.

Bird Origins. The oldest known bird fossil is *Archaeopteryx lithographica* from about 145 million years ago. Scientists consider it a bird because it has feathers. Like modern birds, *Archaeopteryx* shared several traits with some meat-eating dinosaurs. These included the structure of the middle ear and the jaw, the pattern of growth in the long bones, a specialized ankle joint, and hollow bones.

Modern birds are similar to reptiles in other ways as well. Both have scales on rear limbs and feet, for example, and lay their eggs out of water. In addition, birds are genetically closer to reptiles than to any other group of animals. They are probably descended from distant reptile ancestors. Some paleontologists are asking whether those ancestors were dinosaurs or some earlier, still unknown ancestor.

The study of certain fossils found in the 1990s may help answer that question. *Confuciusornis sanctus* from China, perhaps 122 to 145 million years old, may have been the world's first beaked (toothless) bird. Since the main line of bird evolution did not lose teeth and develop horny bills until about 65 million years ago, the fossil either reflects a very rapid evolution of toothlessness or a close connection to the beaked dinosaurs. If *Sinosauropteryx prima,* also from China, proves to have been a feathered dinosaur, it will clearly support the claim that birds are descended from dinosaurs. The 75 million-year old *Rahonavis ostromi,* found in Madagascar, is a primitive bird with a retractable sickle claw that closely resembles the slashing claws of certain dinosaurs, including *Velociraptor.*

Bird Flight. Scientists do not yet know when or how birds began flying. The "trees-down" theory is that the first birds were tree climbers who jumped between branches, used their wings to glide, and gradually developed the ability to flap their wings and fly. The "ground-up" theory is that early birds were runners that used their small forelegs for balance; over time the forelegs evolved into wings, and the flapping, balancing motions developed into flight.

Either way, flight requires a lot of energy and oxygen. Moving beyond the reptile pattern, birds developed a four-chambered heart that supplies oxygen to the muscles more efficiently, making long periods of activity possible. Birds also developed a "one-way" breathing system different from the "back-and-forth" breathing of other land-dwelling **vertebrates.** This one-way system also supports the high oxygen needs of flight-related muscles.

vertebrate any animal that has a backbone or spinal column

Small bird fossil in Senckenberg Museum, Frankfurt, Germany.

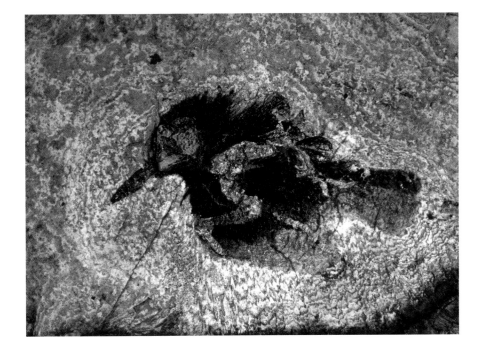

Warm-Blooded Birds.
Modern birds are endothermic, or warm-blooded, which means that they maintain a constant body temperature in spite of changes in their environment. Reptiles are ectothermic, or cold-blooded, meaning that the external temperature affects their body temperature. Endotherms use considerable energy to maintain a steady internal temperature, but warm-bloodedness has advantages. Warm-blooded animals have greater stamina than cold-blooded ones and can remain active in colder environments.

Some evidence suggests that the first birds may have been cold-blooded. Scientists believe, however, that warm-bloodedness evolved in birds shortly after their origin.

Nesting.
As birds became warm-blooded, it became more important to keep their unhatched eggs at fairly steady temperatures. Primitive birds probably buried their eggs to keep them warm, as reptiles do. Shallower soils may have permitted only partial burial of eggs, and birds may have had to shield their eggs from sunlight or keep them warm at night. Brush turkeys in Australia and New Zealand today lay their eggs over warm, decaying vegetation or geothermal heat sources (heat that comes from the Earth's interior) and regulate the temperature by putting material over the eggs or removing it.

Most modern birds, however, incubate their eggs, or keep them warm, by sitting on them. The first birds to incubate eggs by direct contact probably laid the eggs on the ground, as whippoorwills do today. Other species, scientists are sure, developed ways of protecting their eggs from ground-dwelling predators. Modern fairy terns balance their eggs on tree branches. Other modern species of birds lay eggs in natural hollows in rocks and trees, or they create their own hollows. Still other species build nests, ranging from cups hollowed out in the sand to hanging nests of woven vegetation, to hold their eggs.

Parental Contribution.
Wherever they may hatch, the survival of young birds requires a parental contribution. This contribution is the energy that parent birds put into the nest, eggs, and young.

Extinct Bird Giants

Both dinosaurs and birds roamed the Earth during the Cretaceous period, which ended 66.4 million years ago. Like the dinosaurs, many Cretaceous birds became extinct. Among them was an impressive species called *Hesperornis*. This flightless, marine diving bird measured more than 2 m in length. After the Cretaceous extinctions, new kinds of birds evolved. Some became the ancestors of modern birds, while others in turn vanished into extinction. *Diatryma,* for example, has no living descendants. Perhaps this is just as well: it ate meat, stood more than 2 m tall, and had a huge, powerful bill.

a. Giant bird *Hesperornis* was extinct by the end of the Cretaceous period.

b. Another giant, *Diatryma*, was extinct by the end of the Tertiary period.

a.

b.

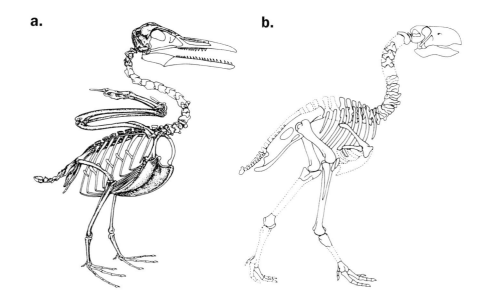

Species vary in the ways in which they contribute that energy. Birds that lay large eggs and incubate them for long periods often hatch young who can walk, swim, or fly very soon and do not need much care. On the other hand, birds who lay small eggs with short incubation periods tend to produce un-feathered, blind, helpless young who must be fed and guarded for a time.

One extreme example is the brush turkey, whose young are quite independent upon hatching and may never see either parent. The wandering albatross, on the other hand, cannot fly until it is about 275 days old and requires extensive parental care. Scientists who believe in the dinosaur-bird link are studying modern bird behavior for clues as to how dinosaurs may have nested and cared for their young. SEE ALSO AMPHIBIANS AND REPTILES; DINOSAURS.

Bjerknes, V.F.K. and J.A.B.

Vilhelm Frimann Koren
1862–1951

Jacob Aall Bonnevie
1897–1975

Norwegian founders of meteorology

The father-and-son team of Vilhelm Frimann Koren Bjerknes and Jacob Aall Bonnevie Bjerknes laid the foundation of modern meteorology, or weather science. Their work allowed scientists to chart the movements of air masses accurately, providing the basis for long-term weather prediction.

The Elder Bjerknes. Vilhelm Bjerknes was born in Oslo, Norway, and absorbed an interest in research from his father, who was a scientist. After studying mathematics and physics, Vilhelm Bjerknes became a professor of science and engineering at the University of Stockholm.

Bjerknes carried on research in hydrodynamics (the forces involved with fluids in motion) and thermodynamics (the properties of heat). His main scientific contribution was to apply his understanding in these fields to the largest fluid system—the world's atmosphere and oceans. Bjerknes and his colleagues studied how friction and the heat of the sun influenced movement in the atmosphere. They believed that scientists could use knowledge of the present conditions of the atmosphere and the oceans to predict future conditions.

Vilhelm Bjerknes founded the Bergen Geophysical Institute in Norway. It became a training ground for many of the world's leading meteorologists of the early 1900s. One of them was Bjerknes's son, Jacob Bjerknes.

The Younger Bjerknes. Jacob Bjerknes shared his father's interest in fluid systems and weather forecasting. From the start of his career, he was a key member of what was known as the Bergen school of meteorology. The younger Bjerknes was the chief of the Bergen Geophysical Institute from 1919 until the 1930s. After leaving Bergen, he moved to the University of California at Los Angeles (UCLA), where he developed a meteorology program. He lived in Los Angeles until his death.

Jacob Bjerknes's early work was the study of similarities in the formation of rain **squalls** and **cyclones.** He borrowed from military language the terms *line* and *front* to describe moving air masses. These terms are now part of every weather reporter's vocabulary. The work of Bjerknes and his colleagues on the movements of the upper and lower atmospheres provided the basic model for the first computer-aided weather forecast in 1950. In addition, Bjerknes's studies of the relationship between atmosphere and ocean led to later work that focused on El Niño and other phenomena of climate and weather. SEE ALSO EL NIÑO AND LA NIÑA; METEOROLOGY.

squall sudden windstorm, often accompanied by rain

cyclone violent, rotational windstorm, such as a tornado

Black Hole

See Stars.

Bowen, Norman Levi
1887–1956
Canadian geologist

petrologist scientist who studies the origin, structure, composition, changes, and classification of rocks

physical chemistry branch of chemistry concerned with the physical properties of materials, such as their electrical or magnetic behavior

Norman Levi Bowen is considered the most outstanding **petrologist** of the 1900s. His importance stems from the many new concepts he introduced and from his application of **physical chemistry** to complex geological problems. His ability to make critical field observations and then test them with simple laboratory experiments evolved into a scientific method of lasting value.

Born in Kingston in Ontario, Canada, Bowen became fascinated by rocks while exploring abandoned quarries as a boy. After earning a degree in chemistry and mineralogy at Queen's University, he did geological field work for the Ontario Bureau of Mines in the summer of 1907. That fall he returned to the university to study geology and mining engineering.

In 1909 Bowen went to the Massachusetts Institute of Technology (MIT) to study under fellow Canadian Reginald Daly. The following year, he became a research student at the Geophysical Laboratory of the Carnegie Institution in Washington, D.C. During the next three years, Bowen helped to survey the proposed route for the Canadian Pacific Railway and led field research in the Frazer River Valley of western Canada.

Between 1912 and 1915 Bowen developed a number of critical ideas in petrology, and publication of *The Later States of the Evolution of the Igneous Rocks* (1915) brought him international recognition. In the years afterward, he published papers on various petrology topics. His studies and field work led to discoveries that had an important impact on such subjects as glass manufacturing and **metallurgy.**

From 1937 to 1946, Bowen worked at the University of Chicago, serving for two years as chairman of the Department of Geology. He then returned to the Geophysical Laboratory in Washington, where he remained until his death in September, 1956. SEE ALSO CAREERS IN THE EARTH SCIENCES; DALY, REGINALD ALDWORTH; PETROLOGY.

metallurgy science that deals with separating metals from their ores and working with metals

Brachiopods

invertebrate animal without a backbone

mollusk marine animal with a shell of one or more pieces that encloses a soft body

GT See Geologic Timescale on page viii.

species narrowest classification, or grouping, of organisms according to their characteristics; members of a species can reproduce only with others of that group

niche specific position or role of a group of closely related organisms

Brachiopods are a group of shelled marine **invertebrates** that have two valves (shells), most commonly made of calcium carbonate, completely enclosing the body and a fleshy tentacle-like structure for gathering food. Although similar in external appearance to clams and other **mollusks,** brachiopods have completely different internal organs and are unrelated to mollusks.

Adult brachiopods are incapable of moving from place to place. Some attach themselves to the seafloor or other objects by means of a short, fleshy stalk called the pedicle. Others lack a pedicle and lie in a stationary position on the seafloor. Brachiopods often provide an attachment surface for small corals and other marine organisms.

Brachiopods range in size from a few millimeters to several centimeters across. Their shells may take a variety of concave (inward-curving) and convex (outward-curving) shapes. The surface features of brachiopods also vary. Some have smooth shells with fine lines; others have coarse ridges and furrows.

Brachiopods first appeared nearly 600 million years ago, in the early Paleozoic era[GT]. At that time, they were among the most abundant animals on the ocean floor, with thousands of different **species** occupying a wide variety of marine environments. Over time the brachiopods declined in number and variety, perhaps as a result of competition from other marine species or from loss of habitat caused by climate change or other factors.

Today there are fewer than 325 species of brachiopods in the world's oceans. Most are found in relatively shallow water and occupy small, hidden **niches** in the marine environment. SEE ALSO INVERTEBRATES; MOLLUSKS.

Bragg, William Henry and William Lawrence

**William Henry
1862–1942**

**William Lawrence
1890–1971**

English physicists

mineralogy science that deals with minerals and their properties

atomic structure number and organization of tiny particles within the atoms that make up all matter

radiation energy emitted as particles or waves

diffraction occurrence in which waves of light or other types of waves bend as they meet obstacles

crystal any solid whose internal structure is arranged in a repeating, orderly pattern

The father-son team of William Henry and William Lawrence Bragg made significant contributions in the fields of physics and **mineralogy.** Together they established the technique of using X rays to determine the basic compositions of solid materials. Working separately, they used this and other methods to reveal the **atomic structures** of some of the most common and important minerals found on Earth.

Born in Westward, England, William Henry Bragg studied mathematics at Cambridge University. After graduating he accepted a professorship at the University of Adelaide in Australia. From 1886 to 1904, Bragg devoted most of his energy to teaching and developing a scientific curriculum for the university. He did not begin research until age 41, when he began to study the ways in which X rays and other forms of **radiation** are absorbed and scattered. His observations led to the development of a method for identifying radioactive substances.

Bragg's eldest son, William Lawrence, was born and educated in Australia, but he returned with his family to England in 1909, when his father was appointed as a professor of physics at the University of Leeds. Like his father, the younger Bragg entered Cambridge University, where he studied mathematics and physics. In 1912 William Lawrence began an X-ray investigation of certain chemical substances and produced a mathematical model for the process of **diffraction.** His studies established that X rays act in a manner similar to that of light waves and proved that **crystals** consist of atoms aligned in regular patterns.

Using these discoveries, the two Braggs worked together and obtained precise values for the X-ray wavelengths emitted by various metals when the metals are bombarded with electrons. These findings and the diffraction model led to experimental determination of the internal atomic arrangements of a number of different minerals. The Braggs' scientific method remains fundamental to the analysis of crystal structures such as minerals. For their work, they were jointly awarded the Nobel Prize in physics in 1915.

From 1935 to his retirement in 1940, William Henry Bragg served as president of the Royal Society of London, one of the oldest scientific organizations in Europe. His many honors included knighthood in 1920 and the Copley medal of the Royal Society in 1930. William Lawrence continued teaching and conducting research until his retirement in 1966. Among the younger Bragg's honors were several medals of the Royal Society, knighthood in 1941, and the Companion of Honor in 1967. SEE ALSO EARTH MATERIALS, CHEMISTRY OF; MINERALS.

Sir William Henry Bragg.

Brongniart, Alexandre
1770–1847
French geologist

mineralogy science that deals with minerals and their properties

sedimentary rock rock formed from deposits of sediments (soils, rocks particles, and other materials) over long periods of time

Alexandre Brongniart made significant contributions to the young science of geology in the early 1800s. An accomplished field geologist and careful observer of nature, he helped to demonstrate the usefulness of fossils in interpreting Earth's history.

Born in Paris, France, Brongniart began studying for a career in medicine or pharmacy but became interested in geology while serving in the military. In 1794 he received an appointment as a mining engineer for the government. He later taught at university and, in 1822, became chairman of **mineralogy** at the Museum of Natural History in Paris.

Brongniart established a close working relationship with French scientist Georges Cuvier. Together they analyzed the geology of the Paris region and made a detailed geologic map of the area. They called attention to breaks, now known as disconformities, that separated different groupings of fossils in **sedimentary** rock layers. They identified differences between rock strata, or layers, based upon whether the layers contained fossils from freshwater or ocean species. Brongniart also demonstrated that rock strata in the area could be correlated with strata elsewhere in Europe.

Brongniart shared with Cuvier a belief in catastrophism—the idea that sudden changes sometimes occur in Earth's development. They considered their observations of disconformities as evidence of this theory, noting that geologic "revolutions" might be responsible for the birth of mountains and extinction of plants and animals. An innovative feature of their work was the recognition of subtle variations of rock type within the same geologic formation. They attributed these differences to varying environmental conditions when the rock layers were formed. SEE ALSO CUVIER, GEORGES; FOSSILIZATION AND THE FOSSIL RECORD; GEOLOGY, HISTORY OF; SEDIMENTS AND SEDIMENTARY ROCKS; STRATIGRAPHY.

Carbon-14 Dating

isotope one of two or more forms of the same element; isotopes differ in mass (the amount of matter that causes an object to have weight in an area influenced by gravity)

organic related to carbon compounds, which are found in living things; living or formerly living

Carbon-14, or C-14, is an unstable (radioactive) **isotope** of carbon formed by cosmic rays in the Earth's atmosphere. Carbon-14 dating, also called radiocarbon dating, is a technique for determining the age of an **organic** object by measuring the amount of C-14 it contains.

All living things use carbon. Plants, for example, take in carbon dioxide (CO_2) during **photosynthesis.** A certain amount of the CO_2 they use will contain C-14. As long as the plant lives, it will maintain a constant level of C-14. Once it dies, however, it ceases to take in additional C-14, and the amount of C-14 it contains will steadily decrease as the radioactive element decays to form stable (nonradioactive) nitrogen-14, or N-14.

Careers in Astronomy and Astrophysics

photosynthesis process by which plants manufacture food from sunlight

half-life amount of time required for half of a given amount of a radioactive substance to decay, becoming a different substance

Years later, scientists can examine the amount of C-14 remaining in a formerly living substance—for example, a lump of coal—and compare this with the amount the substance would have contained when it was living tissue. C-14 has a **half-life** of about 5,568 years, so a fossil substance that contains half the amount of C-14 it would have contained when alive must have died 5,568 years ago. Thus, measuring the C-14 level in any organic substance can yield an approximate date for the time of death.

C-14 dating is most accurate for measuring the ages of objects close in age to the half-life of C-14. It is therefore especially useful in studies of human history and prehistory as well as for dating geologically recent events (300 years to somewhat more than 60,000 years), such as glacial and volcanic activity. It has also been used for certain other purposes, such as studying changes in the Sun. SEE ALSO LIBBY, WILLARD F.

Careers in Astronomy and Astrophysics

astrophysics science that deals with the physical characteristics of stars and stellar events or conditions that can be observed

Astronomy is traditionally defined as the study of space and the various bodies it contains, including the Sun and Moon, stars and planets, comets, asteroids, and galaxies. In the past, astronomers were limited to making observations of the sky with optical telescopes, which use visible radiation (light that the eye can see). Modern astronomers also use a wide variety of other tools and techniques to study the universe by observing other types of radiation, including radio waves, infrared and ultraviolet light, X rays, and gamma rays. Through these methods, they have uncovered a number of exotic objects deep in space—black holes, pulsars, and quasars are only a few. In addition, the possibility of life on other planets has opened up a rich field for study.

Astrophysics is not the same as astronomy, but it is a closely related science. It includes the study and physical interpretation of the processes observed in the universe. Other fields related to astronomy include geology, physics, chemistry, biology, and the atmospheric sciences. Engineering is also related to astronomy because of the role engineers play in creating some of the tools needed to explore the heavens: manned spacecraft, robotic spacecraft, and Earth-based observatories.

Education and Training. A person interested in a career in astronomy or a related field should have a strong background in science and good mathematical skills. The more theoretical an area of astronomical research, the more mathematical knowledge is required to study it adequately. For example, to do research on the subject of general relativity (gravitation theory), skills in the field of higher mathematics known as tensor analysis are essential. Computer skills are also important because of the role computers play in all scientific endeavors.

Generally, careers in astronomy require at least a four-year college degree in a physical science, life or earth sciences, or engineering. To become a research scientist, a graduate degree is necessary, requiring at least another two to five years of study. Most colleges offer introductory courses in astronomy, but there are only a few with undergraduate majors in astronomy or astrophysics. Specific areas of study, such as astrogeophysics (the geology of moons and planets) or astrobiology (the origin and development of life on Earth and elsewhere) are available only at colleges and universities that have active research programs in these areas as well as a degree program in astronomy.

Careers and Employment Opportunities. Most research and discovery in astronomy is the work of research scientists. Their work centers around planning and performing experiments. They design instruments to make measurements and record the information produced by these instruments. They then analyze and interpret the data in an attempt to discover the underlying laws of some observed event or situation and tie them in with facts that are already known. This level of scientific research generally requires a doctorate.

However, the fields of astronomy and astrophysics are not restricted to astronomers. There are many other occupations that are related to the study and exploration of space. Engineers, for example, help build and operate the instruments used in astronomical research, through which discoveries are

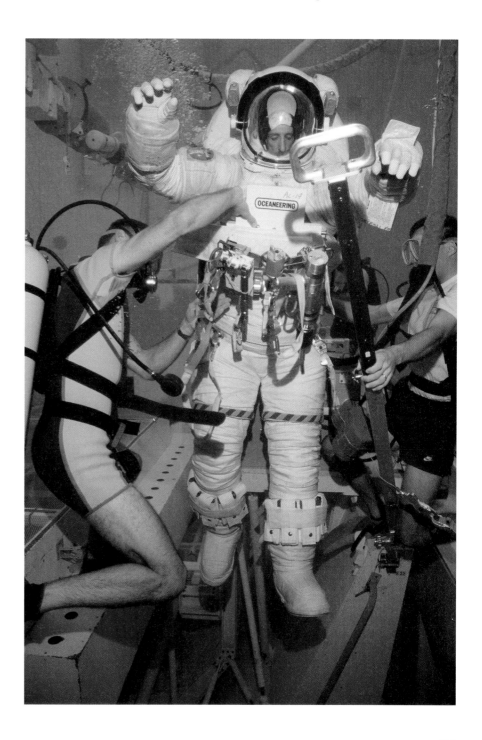

Divers assist an astronaut with his spacesuit in an underwater training facility.

Astronauts experience weightlessness during their training.

made. They may work in the control rooms of many astronomy-related missions, repair computers used to analyze data, and design and build astronomical observatories. Technicians support the work of scientists and engineers and also help in the design, development, testing, operation, and repair of instruments and spacecraft. In addition, the work of those in other branches of science—such as physics, chemistry, biology, and geology—can directly or indirectly support astronomical research.

Many astronomers work for various government organizations. The National Aeronautics and Space Administration (NASA) is one such employer, although NASA's focus is shifting somewhat toward overall management of the space program and toward supporting rather than conducting research. Other government agencies that hire astronomers include the U.S. Geological Survey (USGS) and the National Oceanic and Atmospheric Administration (NOAA). Astronomers may also be employed by government-supported research institutes, such as the Hubble Space Telescope Science Institute and the Lunar and Planetary Institute, or contracting institutions such as the Jet Propulsion Laboratory (JPL). Universities may hire astronomers with doctoral degrees to conduct research, and private industries hire all levels of engineers, technicians, and scientists.

It is hard to know what the future of employment in astronomy and astrophysics will be. Each year, there are only about 100 doctorates given in astronomy, yet there are usually many more astronomers than there are job openings. On the other hand, more positions will open up as astronomers, scientists, and engineers retire. Astronomers may also find work in the other branches of science that support astronomical research but are not restricted to pure astronomy. SEE ALSO CAREERS IN THE EARTH SCIENCES.

Careers in Atmospheric Sciences

The television weather forecaster's job is the most familiar career in the atmospheric sciences, but this field offers many other opportunities, from monitoring air pollution to studying changes in the Earth's climate. At the same time, weather remains central to the atmospheric sciences, and many of those who enter these professions do so because they are fascinated by weather.

Scope of Activities. Meteorology is the science of the atmosphere. Meteorologists seek to understand how the atmosphere works so that they can predict its future state, or forecast the weather. The term "atmospheric science" is often used to describe a combination of meteorology and other physical sciences, such as chemistry. Because people who forecast the weather are often called meteorologists, those who work in related professions—perhaps studying the chemical reactions that take place 45 to 50 km above the Earth—may prefer to be called atmospheric scientists.

National Aeronautics and Space Administration's research airplane *Perseus* is used to collect data on atmospheric conditions.

A scientist releases a weather balloon in Antarctica.

A large percentage of the meteorologists in the United States work for the government, either in the military or in the National Weather Service (NWS). Their primary responsibilities are forecasting day-to-day weather and issuing warnings of severe weather. All meteorologists, including those on television, use information from the NWS and military observation systems as well as the NWS computer models.

The government is not the only source of jobs for those who want to forecast weather. Meteorologists may work for private firms, making forecasts that are tailored to the special needs of customers such as power utility companies. Private firms also supply weather forecasts to many radio stations.

Some atmospheric scientists and meteorologists conduct research at government laboratories or university research centers. Their work may involve the study of global issues, such as the effects of atmospheric pollution on the Earth's climates, or it may focus on a local or highly specific project, such as the airflow inside a tornado. University professors of atmospheric sciences also do research in addition to their teaching duties.

Educational Requirements. Almost all of these career opportunities require at least a bachelor's degree in meteorology or atmospheric science. More than 60 colleges and universities in the United States have programs leading to such degrees. The American Meteorological Society offers some college scholarships to students interested in careers in the atmospheric sciences.

College students take courses in **calculus,** physics, and chemistry during their first two years, before beginning the more specialized courses in meteorology or atmospheric science. High school students interested in the atmospheric sciences should prepare themselves by taking as many mathematics and science courses as possible before entering college.

Many scientific career opportunities—especially teaching at the college level or conducting research—require a master's degree or doctorate. Students who do very well as undergraduates may be encouraged to pursue a graduate degree. Most schools provide financial assistance for graduate students.

Career Prospects. Atmospheric science is a relatively small field in the United States, although it is growing. New graduates can generally find jobs,

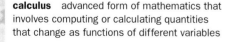

calculus advanced form of mathematics that involves computing or calculating quantities that change as functions of different variables

but they may not have the luxury of choosing among several job offers or of getting the exact job of their choice.

During the 1990s, the NWS underwent extensive modernization and restructuring so that it could make the best possible use of the latest technology. This program created many new employment possibilities at the NWS.

Another growing area of opportunities for atmospheric scientists and meteorologists is in private consulting firms. Clients such as city governments and corporations hire consulting meteorologists to perform various tasks, such as carrying out the environmental impact studies and pollution control activities required by Clean Air Act and other laws.

Society's increasing concern with the environment, especially with such global issues as the destruction of the ozone layer and the possibility of climate change, means an increase in government funding of research projects connected with these issues. Such projects offer new career possibilities for aspiring atmospheric scientists. SEE ALSO ATMOSPHERE, EARTH'S; METEOROLOGY.

Careers in Environmental Science

deforestation   clearing trees from an area without replacing them

erosion   wearing away of land by wind and water

global warming   theory that suggests a gradual warming of the Earth's atmosphere as a result of increasing air pollution

hydrogeology   branch of geology that deals with water in relation to Earth structures

environmental engineering   branch of engineering that specializes in dealing with environmental problems

geochemistry   science that deals with the chemical composition of the Earth's materials

Environmental science is a wide-ranging and complex field, concerned with the study of the natural environment and the effects upon it of human activity and natural events. The field deals with such issues as **deforestation**, soil **erosion**, air and water pollution, **global warming**, and the destruction of plant and animal habitats.

Because its concerns are so broad and because many environmental problems are interconnected, environmental science encompasses a number of varied areas of science. In addition to broader disciplines such as biology and geology, it also includes specialized fields, such as **hydrogeology**, **environmental engineering**, and **geochemistry**.

Development of the Field.   Environmental science as a field of study first arose in the 1960s, largely in response to increased environmental concerns in the United States and elsewhere. The creation of the U.S. Environmental Protection Agency (EPA) in 1970 spurred further development of the field. The EPA is a federal agency whose mission is to protect the environment. Its responsibilities include monitoring pollution and studying its effects.

The environmental movement gained strength in the 1970s and 1980s as public interest in conservation grew. During those years, federal, state, and local governments passed many laws to clean up and protect the environment. These included the Superfund Act, to deal with toxic wastes, and the Clean Air Act and Clean Water Act, which focused on improving the quality of the nation's air and water. Another law, the National Environmental Policy Act, required federal agencies to prepare an environmental impact statement before building any structure, such as a dam or highway, that might affect the environment. The impact statement assesses the potential for environmental harm from such projects to determine whether or not they should be carried out.

The growing interest in environmental issues in the 1970s and 1980s, along with the new laws, created a demand for specialists with knowledge and training in these matters. As a result, the number of jobs for environmental scientists expanded dramatically.

Education and Training.   The complexity of environmental science makes it impossible to detail a single course of study. A strong foundation in science is always essential, but students can choose to focus on biology,

Careers in Environmental Science

A scientist tests sample of groundwater for pollution from nearby landfills. Test results appear on the screen of a laptop computer.

chemistry, geology, physics, or other subjects, depending on their interests and career goals. Some colleges offer environmental science degrees, which usually combine study in several science areas. Mathematics and computer skills are very important.

Individuals who want to specialize in a particular branch of environmental science focus their study on academic subjects directly related to it. For example, those who are interested in the substances that contribute to air and water pollution would concentrate their studies on chemistry. Those who want to study the hazards of volcanoes, earthquakes, landslides, and floods need to emphasize geology. Within each of these general areas, students might also focus on specialized subjects such as **biochemistry**, hydrogeology, or atmospheric science.

A bachelor's degree in the geosciences would meet the qualifications for some entry-level jobs in environmental science. Further study—a master's degree or doctorate—is needed for most higher level work in the field. Concentration in a specialized field of study and skill in using sophisticated technology also help environmental scientists advance their careers.

Careers and Employment Opportunities. There is great diversity among careers in environmental science. In addition to the many branches of study that make up the field, there is a variety of employers, including schools, colleges, government, private industry, and nonprofit organizations. While many specialists work in research, others concentrate on education, engineering, politics, wildlife or forest management, or other areas. Some environmental scientists work mostly in offices or laboratories. Others travel to various areas of the world to gather data that will help them answer particular questions, such as how to contain hazardous wastes safely and effectively.

Many environmental scientists work for federal, state, and local government agencies as researchers, inspectors, and administrators. These people often have an opportunity to guide public policy. They also help start programs to protect natural resources—such as water and soil—and to dispose of wastes. Environmental scientists play a major role in helping communities to prepare for and cope with hazards such as volcanoes, earthquakes, landslides,

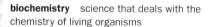

biochemistry   science that deals with the chemistry of living organisms

and floods. For example, they may produce maps and develop procedures to minimize harm to people and property.

Environmental scientists employed by private industry often work to reduce pollution in order to meet the requirements of federal, state, and local laws. They may also develop products that will protect the environment, such as devices to limit the pollutants released into the air by automobiles or factories. Some scientists find work at environmental consulting firms, providing environmental services to small companies that cannot afford to employ full-time scientists of their own.

Large numbers of environmental scientists work at colleges and universities, where they teach and conduct scientific research. Their research helps to increase knowledge about environmental issues and identify possible solutions.

Employment Outlook.   Despite some successes in improving and protecting the environment, these issues will remain a concern for many years to come. As populations grow, pressures on the natural world will increase. Environmental scientists will therefore be needed to solve not only current problems but also new ones that will arise in the future. Job opportunities in private industry will probably remain plentiful because of the need to assess the environmental impact of industrial activities and prevent damage to natural resources. SEE ALSO CAREERS IN ATMOSPHERIC SCIENCES; CAREERS IN HYDROLOGY; CAREERS IN OCEANOGRAPHY; CAREERS IN OIL AND GAS; CAREERS IN TEACHING IN THE EARTH SCIENCES; CAREERS IN THE EARTH SCIENCES.

Careers in Exploration and Mining

extraction removal from the Earth

Because the discovery, **extraction,** and use of minerals and fuels are vital to the economic well-being of countries and societies, the demand for fuel and mineral resources creates many career opportunities for earth scientists. Some are exploration geologists, who search for new sources of useful natural materials. Others are mining geologists, who determine the best ways of mining or extracting resources. Both fields are changing in response to new needs, new environmental concerns, and new technology.

Mining engineers examine core samples at a site being evaluated for platinum mining.

Careers in Exploration and Mining

mineralogy science that deals with minerals and their properties

extract to separate a substance from its source

ore metal-bearing mineral or rock

Scope of Activities. Exploration geologists search the Earth's crust for deposits of valuable elements and minerals. They also evaluate the chemical and physical properties of known deposits. Exploration geologists may look for concentrated sources of high-value metals such as gold and platinum, or they may look for less costly but very useful materials such as clay, limestone, or sand and gravel. Some search for mineral fuels, including uranium and coal.

Exploration geologists may perform large-scale area surveys followed by evaluations of smaller specific sites. Site evaluation calls for many basic geological studies and skills, such as **mineralogy** and mapping. Geologists may use physical measurements and chemical analyses to help them determine what minerals lie below the surface of the Earth. Those who apply the principles of chemistry to their work are called geochemists, while those who combine geology with physics are called geophysicists.

Mining geologists provide information to engineers that helps them to **extract** grades of **ore** that offer the best return for the cost of recovering the desired minerals or metals. This activity is called grade control. Mining geologists also examine mined material to see whether it contains any additional resources, and they work with engineers to solve practical problems such as those concerning mining safety and waste disposal.

Educational Requirements. The discovery of natural resources will always require educated and experienced people. Earth scientists must have a fundamental background in mathematics and the basic sciences to stay current with new techniques and knowledge. Exploration geologists must also have a thorough knowledge of many aspects of geology so that they can work with rocks of all ages and types.

An exploration geologist needs broad undergraduate training followed by specialized education in graduate school. Experience gained from observations in the field is equally important. Familiarity with other languages is also an advantage, both for occasional field work in other parts of the world and because job opportunities are growing most rapidly outside the United States.

Exploration and mining geologists, who once worked only with their employers, now interact with regulatory agencies, environmental groups, and government administrators. These professions remain scientifically and technically demanding, but good communication and interpersonal skills have become important assets.

Employment Trends. Both governments and private companies employ exploration and mining geologists. Resource exploration is incomplete in many parts of the world because of political situations or because the areas are inaccessible. However, exploration involving the ground search of these regions continued during the 1990s and will probably continue far into the future.

The future of geological exploration within the United States, however, is less certain. National debates concerning land ownership, environmental issues, and the value of mineral resources have influenced decisions about land use. Although American industry and government will continue to draw upon the specialized knowledge and skills of exploration geologists in solving resource-related problems, career opportunities are increasing most rapidly in other countries.

Technology is also changing the ways both mineral exploration and mining are carried out. Skilled and experienced scientists will always be needed to evaluate materials in the field, but exploration geologists are making greater use of tools such as remote imaging (using satellite or aerial photography) and computer modeling to enhance their knowledge of distant or unexplored places. With an eye to the future, mining geologists are focusing on new methods of extracting minerals and reducing the environmental impact of mining. **SEE ALSO** ECONOMIC GEOLOGY; MINERALOGY.

Careers in Hydrology

surface water water found above ground in lakes, ponds, rivers, and streams

groundwater underground water that supplies natural springs and can be tapped by wells

geochemistry science that deals with the chemical composition of the Earth's materials

Hydrology is the study of the occurrence, distribution, movement, and quality of **surface water** and **groundwater.** It incorporates various other disciplines, including fluid mechanics, biology, chemistry, geology, mathematics, and physics. Hydrology is of huge importance today because of the increasing needs for water in agriculture and industry as well as for human consumption.

Large quantities of usable water are needed to support modern, industrialized societies. Some places, however, have shortages of water. Moreover, human activities often contaminate available water supplies. Lack of good-quality water can cause serious health problems, affect food production, and hinder economic development. Hydrologists seek solutions to such problems.

The Work of Hydrologists.　Hydrologists examine all aspects of water, including its physical, biological, and chemical characteristics. Their work involves such activities as geologic mapping and surveying, well drilling, taking water samples and measurements, determining hydrologic and geologic boundaries, designing computer models of water systems, and proposing action to protect water supplies.

Training and Education.　To work as a hydrologist, an individual must have at least a bachelor's degree in physical or natural science or in engineering. Individuals with a master's degree in hydrology or a related field generally have a better chance of employment.

Many universities and scientific societies now offer short courses, seminars, and training programs in hydrology. In addition, academic degrees are available in related sciences such as environmental geology, environmental engineering, and environmental **geochemistry.**

Need for Hydrologists.　The contamination of water supplies is one of the most significant environmental issues facing the United States and is the focus of much public concern. Contamination has caused the closing of thousands of public and private wells. Federal and state agencies and private industries will spend large sums of money in the years to come in an effort to clean up contaminated water supplies.

Maintaining good water supplies will require many hydrologists. There is now a shortage of trained hydrologists in the United States. It is estimated that about 10,000 hydrologists are needed to deal with the protection and cleanup of contaminated groundwater and surface water alone.

Employment Opportunities and Outlook.　The employment outlook for hydrologists is very good. Hydrologists are employed in federal, state, and local governments as well as in education and private industry. Funding is available at all levels of government for dealing with water quality and water supply issues. Interest in hydrology has been spurred by activities of the Environmental Protection Agency (EPA). The Department of Defense is involved in cleaning up groundwater pollution at military installations, while the Department of Energy has a similar program to deal with hazardous waste

A hydrologist collects a sample of underground water for testing.

sites. Other government agencies that employ hydrologists include the U.S. Geological Survey, the Bureau of Reclamation, the Soil Conservation Service, the Bureau of Land Management, and the National Park Service.

In private industry, environmental or engineering firms often employ hydrologists. The work generally includes providing assistance to government agencies, assessing industrial sites, and investigating water supplies.

Many hydrologists join professional societies as a way to maintain contact with others in their field. Such societies include the Geological Society of America, the National Ground Water Association, the International Association of Hydrogeologists, and the American Institute of Hydrology. SEE ALSO POLLUTION OF STREAMS, LAKES, AND GROUNDWATER; WATER QUALITY; WATER SUPPLY AND MANAGEMENT.

Careers in Oceanography

More than two-thirds of the Earth's surface is covered with oceans. Oceanography is the scientific study of the oceans, and it offers a wide variety of career possibilities.

Development of the Field. Oceanography began with the British-supported *Challenger* Expedition (1872–1876), which systematically explored all the major oceans. Although whalers, sailors, and other seafarers knew much about the oceans, their knowledge was guarded rather than shared with others. For example, logbooks of early Portugese and Spanish explorers were considered state secrets.

As scientists began to share information in the late 1800s, **fisheries** offered the first employment opportunities for people who were interested in the ocean. Some countries set up organizations to study their fisheries, find new resources, and protect existing resources.

The field of oceanography began to take shape as the demand for ocean scientists grew, and universities started to offer training programs in oceanographic studies. During World War II (1939–1945), ocean scientists became

fisheries businesses involved in the harvesting of marine animals and the places where such animals are found

important in naval warfare; they performed duties such as helping the military determine where to land troops on seacoasts. After the war, American universities began to offer courses and degree programs that trained researchers and scientists for military tasks such as tracking submarines. Later the focus of oceanography shifted to the protection and regulation of fisheries and an improved understanding of the role of the ocean in climate change and its effects on the Earth.

Educational Requirements. Students who want to work in oceanography must take a broad variety of science courses at the undergraduate or graduate level. The study of the ocean involves many physical sciences such as physics, **meteorology,** geology, and **hydrology** as well as life sciences such as marine biology.

Until the 1990s, most universities advised prospective oceanographers to obtain a bachelor's degree in a traditional field of science—perhaps biology or physics—and then study oceanography at the graduate level. However, more American universities have begun offering undergraduate programs leading to bachelor of science degrees in oceanography because the job opportunities for graduates with B.S. degrees are increasing. Even so, a significant percentage of American oceanographers have doctorates and teach or conduct research at the university level.

Career Opportunities and Trends. Changes in global politics during the 1990s brought a reduction of funds for defense and national security activities, including the military applications of oceanography. Fisheries, however, are still a major employment area, and several universities have programs for fisheries scientists. Many countries hire oceanographers to prepare maps for commercial fishing fleets, showing where certain fish can be caught. Oceanographers also supply similar products and services to those who fish for recreation.

meteorology science that deals with the atmosphere, especially with regard to climate and weather

hydrology science that deals with the occurrence, distribution, movement, and quality of water

Underwater Pioneer

Although Jacques-Yves Cousteau (1910–1997) did not receive formal training in oceanography, he became the world's best-known figure in underwater exploration and research. Together with associates in France, Cousteau invented the Aqua-Lung, the first device for breathing underwater without a hose connected to an air source above water. By 1946 the Aqua-Lung had made scuba diving possible for people around the world. Cousteau later pioneered underwater filmmaking. Millions enjoyed and learned from his movies and television series. Although the authenticity of some of Cousteau's underwater scenes was sometimes questioned, his dedication to drawing public attention to undersea life and the need to protect it makes him an important figure in oceanography.

Fisheries scientists examine bottom samples collected from waters 80 m deep.

aquaculture practice of growing marine plants and raising marine animals for food

Aquaculture is another emerging area of employment for those who are trained in oceanography. As wild fish stocks are overexploited and become contaminated by polluted ocean environments, the business of "farming" fish, shrimp, and other sea organisms is rapidly growing. China and Japan already have multibillion dollar aquaculture industries. Aquaculture is expanding to Europe and North America, creating a need for people trained in oceanography and related fields.

The petroleum industry employs numerous oceanographers in the development of offshore oil and gas resources. Many countries, such as Norway and Great Britain, produce nearly all their oil and gas in offshore fields, where deposits are located on **continental margins.** Because these areas provide the most promising locations for new oil and gas deposits, the worldwide oil and gas industry is likely to remain an important area of employment for oceanography.

continental margin region where continental crust meets oceanic crust

Many oceanographers work to protect the environment and are employed by government regulatory agencies at the local, state, and federal levels. These marine scientists also do research and help to restore damaged environments such as wetlands.

Remote sensing from satellites provides oceanographers with enormous amounts of information to use in research and monitoring the seas. At the same time, increasingly powerful computers allow that data to be stored and applied. Oceanographers are finding new employment opportunities as computer experts and data managers. SEE ALSO CAREERS IN OIL AND GAS; OCEANOGRAPHY; OCEANOGRAPHY, HISTORY OF.

Careers in Oil and Gas

Oil and natural gas are among the world's most important natural resources, and they play a crucial role in the economies of modern nations. Oil and other petroleum products are used to produce fuels, plastics, synthetic fibers, lubricants, and various other chemicals. Natural gas is an efficient source of energy commonly used for cooking, heating, and generating electricity. Historically, most of the earth scientists in the United States have been involved in finding, developing, and producing petroleum and natural gas.

Careers in oil and gas exploration combine scientific discipline with the excitement of discovery. Earth scientists, primarily in the fields of geology and **geophysics,** commonly use extensive scientific data to help them decide where to drill for oil and gas. Because they must also use their imaginations and make educated guesses about the possible locations of deposits, earth scientists who explore for oil and gas are often called "wildcatters." If wildcatters are wrong, the well is dry, but if they are right, the rewards are immediate.

geophysics science that deals with the physical characteristics and structure of the Earth

Education and Training. In the early days of oil and gas exploration, most earth scientists working in the industry held bachelor's degrees, chiefly in geology. Since the late 1950s, however, graduate training has become more important. Today most oil and gas geologists hold advanced degrees.

All earth scientists who engage in the exploration and development of oil and gas sources must have strong backgrounds not only in geology but also in mathematics, physics, and chemistry. Many petroleum engineers have additional training in the technical aspects of oil and gas **extraction.** As geologists and geophysicists become more involved in oil and gas production and in advanced recovery processes, it becomes increasingly important for them to understand basic engineering principles. Because oil and gas affect the national economy, earth scientists who wish to work in higher-level management in

extraction removal from the Earth

Petroleum engineers test the oil pipe and drill of an oil well.

the industry should also have training in economics, law, business, and public policy.

Employment Outlook. The number of earth scientists working in the oil and gas industry has decreased since the early 1980s, when the United States experienced an oil "boom." The increased use of technology in locating and extracting oil and gas has also reduced the overall need for earth scientists in the field. Despite the reduced level of employment in the oil and gas industry, opportunities (both in the United States and abroad) will probably remain fairly steady for many years to come. The stability of career opportunities is heavily dependent on the remaining deposits of gas and oil.

The United States can supply its own expected demand for natural gas and at least half of its oil needs up to about the year 2050. Scientists estimate that more than 6 trillion barrels of oil remain unrecovered or undiscovered throughout the world. Vast amounts of natural gas also remain untapped. The extent of the remaining oil and gas reserves around the world means that earth scientists will continue to be involved in exploration and development of these resources for at least the next 100 years.

However, oil and gas resources are nonrenewable—once used, they cannot be replaced—and some will be increasingly difficult and expensive to recover. This situation will change the nature and scope of the work of earth scientists as new technologies are developed that make locating and recovering smaller, less accessible reserves economically feasible. Rewarding opportunities will exist for earth scientists who are able to understand and use these new technologies. SEE ALSO FOSSIL FUELS; OIL AND GAS: CHARACTERISTICS OF; OIL AND GAS: RESERVES, PROSPECTING FOR, RECOVERY OF; PETROLEUM.

Careers in Planetary Sciences

meteorite solid body that has fallen to Earth from space

Although earth sciences generally involve Earth-bound fields of study, a branch known as planetary sciences is concerned with the study of other worlds. Planetary scientists gain much of their knowledge from observations made from Earth, from **meteorites,** and from data—including samples—brought back by manned and unmanned space missions. Although planetary sciences is not a large or rapidly growing field, it does offer some unique career opportunities for those interested in worlds beyond their home planet.

Growth of the Field.   Before the 1960s, a few pioneering scientists examined meteorites and studied the Moon and the planets through telescopes. Their debates centered upon such questions as the origin of the Moon, the source of meteorites, and whether volcanoes or meteor impacts had caused the craters that cover the Moon's surface.

In the 1960s, the United States became determined to land astronauts on the Moon. At that time, planetary geoscience became a significant area of work for earth scientists. Planetary scientists tackled new questions, established a way to assign relative age to lunar materials, and prepared to interpret the geologic history of another planetary body—the Moon.

The beginning of the space age saw the emergence of a new area of geology in the U.S. Geological Survey, the Branch of Astrogeology. The Jet Propulsion Laboratory, run for the National Aeronautics and Space Administration (NASA) and other government agencies by the California Institute of Technology, concentrated on unmanned missions, leaving manned missions to the Manned Spacecraft Center (later the Lyndon B. Johnson Space Center of NASA).

When the first samples of lunar rocks and soil returned to Earth with astronauts in 1969, geologists from around the world helped analyze them. Then in the 1970s and 1980s, many successful unmanned missions to the planets resulted in a huge increase in the amount of information available to planetary scientists. However, the number of planetary missions dropped in the 1980s, causing a decline in the number of research jobs available in planetary science.

Education.   Planetary science is primarily a graduate-level field. Because the field is fairly small, most graduate advisors suggest that students specialize in more traditional Earth-bound topics. A few schools, however, now include

Three shuttle astronauts examine equipment inside a space shuttle flight simulator.

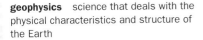

geophysics science that deals with the physical characteristics and structure of the Earth

planetary sciences among their study topics. Most planetary science researchers have doctoral degrees, usually in geology and **geophysics** and sometimes in related fields such as astronomy or physics.

Employers and Career Possibilities. Not all students who are interested in the planetary sciences complete their doctorates. Those whose education ends with a bachelor's or master's degree often find jobs that deal with science—although not always with the planetary sciences. For example, they may find careers in teaching at the high school level, writing about science, or science editing. The exploration of the solar system is a popular topic for magazines and other media, and each new planetary encounter—such as the Mars Pathfinder mission of 1997—revives public interest.

Research careers in the planetary sciences are concentrated in government and university laboratories and in research institutions. There are also jobs in the aerospace industry, which designs and manufactures equipment for space exploration. These positions require engineering experience.

NASA employs doctorate-level planetary scientists to analyze data from past space missions and to propose new experiments for future missions. The Johnson Space Center and Goddard Spaceflight Center are the primary locations for such work. The Jet Propulsion Laboratory has a group of planetary scientists involved in research that focuses on engineering and the operation of unmanned missions.

Except for a few privately funded research institutions and universities, money for planetary sciences in the United States comes through NASA, and NASA's fortunes are affected by politics. The amount of money that goes into planetary exploration—and therefore the availability of jobs in the planetary sciences—depends upon decisions made by NASA administrators, upon the U.S. Congress, and upon the overall state of the economy and the public's support for space missions.

Students considering careers in the planetary sciences would do well to plan their educations with Earth-bound work areas and alternative jobs in mind. A background in geology and astronomy, for example, could lead to work either in the planetary sciences or in Earth-bound aspects of the earth sciences. SEE ALSO GEOPHYSICS; PLANETARY GEOSCIENCE; SHOEMAKER, EUGENE MARK.

Careers in Teaching in the Earth Sciences

oceanography science that deals with oceans

meteorology science that deals with the atmosphere, especially with regard to climate and weather

global warming theory that suggests a gradual warming of the Earth's atmosphere as a result of increasing air pollution

Earth science includes geology, **oceanography**, **meteorology**, space or planetary sciences, and astronomy. It is concerned with the study of the composition, structure, and evolution of Earth, the solar system, and the universe. The enormous expanse of both space and geologic time entails fundamental concepts not traditionally emphasized in other scientific fields.

Because earth science involves several scientific disciplines, it provides the broad background needed to consider environmental issues such as **global warming** and air and water pollution. It also develops an awareness and understanding of natural phenomena such as earthquakes, volcanic eruptions, floods, and hurricanes and provides a framework for attempts to lessen their impact on human society.

Teaching in Grades K-12

The roles and methods of teachers in the earth sciences have changed dramatically, especially in high schools. In the past, teachers relied primarily on formal lectures or presentations in teaching a subject. Modern classes, on the other hand, are characterized by more student participation in interactive discussions and other activities.

Teachers' day-to-day activities are varied. They include planning and executing lessons, exercises, demonstrations, and field trips; preparing and grading tests; assessing students' performance; preparing grades for report cards; supervising study halls, homerooms, and extracurricular activities; and meeting with parents and school staff to discuss students' performance and other issues.

Effective teachers vary in their methods of presentation and organization and in their styles of interaction with students. However, each will know the subject matter well, be a skillful communicator, demonstrate a strong interest in the students' progress, and have an earnest desire to teach and work with young people.

Education, Advancement, and Salary. Because earth science involves geology, oceanography, meteorology, planetary sciences, and astronomy, a person who wishes to teach should take basic courses in each area and earn a bachelor's degree in one of them. Prospective teachers should also be knowledgeable about teaching techniques and the psychology of learning.

The requirements for certification as a teacher vary from state to state. In general, colleges that offer degrees in education have programs that qualify graduates for certification in the same state. These programs usually include specific education courses and student teaching experience under the supervision of a certified teacher. Most states also require teachers to receive a satisfactory grade on the Praxis test or a similar comprehensive examination.

The National Science Teachers Association (NSTA) offers certification to science teachers who have successfully completed three years of teaching in a school. NSTA certification for earth science teachers also requires completion of a minimum number of hours in specific earth, space, physical, and life science courses; knowledge of mathematics; and evidence of study that links the earth sciences to historical, technological, and societal issues.

Completion of an academic degree and certification requirements are one phase of the educational experience of teachers. An equally important phase involves continuing education through courses and workshops. Most K-12 school systems require teachers to earn additional academic credits throughout their professional careers, and they link these requirements to promotion

and salary increases. Professional associations, such as the National Association of Geology Teachers and the National Earth Science Teachers Association play an important role in continuing education.

Experienced teachers today earn an average of nearly $50,000 per year. Starting salaries of beginning teachers are, of course, much lower. In general, teachers in urban and suburban schools earn more than teachers in rural schools, and public school teachers usually earn more that those in private schools.

Employment Opportunities.   There is a great need today for qualified earth science teachers. The demand may decrease somewhat over time because of intensive efforts in the 1990s to upgrade and enlarge earth science education programs. Nevertheless, population growth and increasing dependence on science and technology will ensure a strong demand for K-12 earth science teachers through the year 2005. These teachers will face the challenge of creating curricula that satisfy the changing needs of their students and help them develop skills they will need in the future.

Teaching in Colleges and Universities

At the college and university level, the term *earth science* covers a number of subdisciplines, including geology, **mineralogy, petrology, paleontology, geochemistry,** and **geophysics.** Experts in these sciences have played a leading role in expanding our knowledge of the world, dealing with environmental problems, and exploring for and **extracting** valuable natural resources.

Responsibilities.   Academic activities in colleges and universities commonly include teaching, research, advising students, attending departmental meetings, and working with other faculty members. Many college teachers, especially in smaller colleges, spend most of their time conducting classes; preparing laboratory exercises, experiments, and field trips; preparing and grading assignments and examinations; evaluating research papers; and advising students. Some also spend a great deal of time conducting research on a particular aspect of earth science, sometimes aided by student assistants. Research is generally more important at large universities with graduate programs, and advancement often depends on producing research papers and reports. Some positions at large universities involve only research and carry no teaching responsibilities.

One of the principal attractions of a career in college teaching is the freedom to organize activities. Faculty routinely decide when they will teach their courses, and most choose the focus of their research. Aside from conducting classes, maintaining regular office hours for meeting with students, and attending faculty and committee meetings, they have significant freedom in making use of their time.

Preparation and Employment.   Those who seek careers at the college or university level must have advanced academic degrees. The minimum qualification is a master's degree, but a doctorate is essential to obtain a position at a strong four-year college or large university. It is becoming increasingly important to have two or more additional years of study and research beyond the doctorate.

Earning advanced degrees in the earth sciences requires many hours of academic course work, field study, and laboratory work; original research;

mineralogy   science that deals with minerals and their properties

petrology   branch of geology that deals with the origin, structure, composition, changes, and classification of rocks

paleontology   science that deals with prehistoric plants and animals through the study of fossils and other remains

geochemistry   science that deals with the chemical composition of the Earth's materials

geophysics   science that deals with the physical characteristics and structure of the Earth

extract   to separate a substance from its source

and proven knowledge and expertise in a chosen subject. A master's degree generally takes a minimum of two years to earn, while a doctorate requires three to six additional years.

Teachers at colleges and universities serve at various levels, from instructor to full professor, and they can receive salaries ranging up to $70,000 or more. Promotion and salary increases come with experience and as a result of contributions to research. A major goal and challenge of faculty is to gain tenure, a status that provides employment security. Faculty members who receive tenure must have outstanding records in teaching, research, and other contributions to their field.

Employment opportunities in the earth sciences at the college level are quite good, and the number of teaching positions is expected to increase at a slow but steady rate through the year 2005. Additional employment opportunities will become available over this period as older teachers retire.

Careers in the Earth Sciences

oceanography science that deals with oceans

database collection of information (especially in electronic form) on a particular topic, organized for easy searching and retrieval

extraction removal from the Earth

hydrology science that deals with the occurrence, distribution, movement, and quality of water

The earth sciences are a broad and rapidly changing field. Because the term *earth sciences* describes any application of science to the study of the Earth, it includes many disciplines—such as geology, **oceanography**, and environmental science—and touches on many others, such as astronomy, physics, and biology. Modern technology has allowed scientists to create and use large **databases** that incorporate information from the various branches of earth science. With the help of these databases, students of the earth sciences can examine the big picture of Earth and its many interacting systems.

Career opportunities in the earth sciences are undergoing major changes in the United States and in other industrialized countries. For example, there has been a marked reduction in jobs related to resource **extraction**, such as mining and oil drilling. However, other areas of the earth sciences, such as environmental management, are expanding rapidly as society grows more concerned with its own impact on the global environment and how best to preserve Earth's resources.

Education and Training. Preparation for a career in earth sciences begins with a strong background in science and mathematics. Undergraduate courses in the fundamental sciences (physics, chemistry, and biology) and higher-level mathematics, such as calculus, provide a basis for more extended work in any of the many branches of earth science. Courses in basic engineering are also useful for students who are interested in fields such as **hydrology**, engineering geology, or exploration and mining.

Careers and Employment Opportunities. Because the field of earth sciences is so large and diverse, earth scientists may find work in almost any setting: private industry, nonprofit organizations, educational institutions, or governmental agencies. However, specific areas of earth science may be better suited to certain types of work environments. For example, earth scientists may find work with private businesses in the fields of resource extraction or leisure activities (such as museums, parks, tours, or cruises). In local, state, and federal government, earth scientists may evaluate natural hazards and help plan for hazard reduction and emergency response, or they may help to shape policies concerning land use and the environment. In addition, it is possible to build a career that combines earth sciences and another field, such as teaching, the arts, medicine, or law. SEE ALSO CAREERS IN ASTRONOMY AND ASTROPHYSICS; CAREERS IN ATMOSPHERIC SCIENCES; CAREERS IN ENVIRONMENTAL

Careers in Writing, Photography and Filmmaking in the Earth Sciences

Careers in earth sciences are not limited to teaching, research, and industrial work. Careers are also available in writing, photography, and filmmaking. For instance, individuals who work in these creative fields write articles or take photographs on science topics for magazines, newspapers, and books, and are involved in making science-related programs for television. They also work for companies that develop films, CD-ROMs, or other science educational materials for classrooms, special clients, or the general public.

A photographer installs remote control cameras near a launch pad at Tanegashima Space Center in Japan.

Skills and Education. The skills needed for a career in writing, photography, or filmmaking vary, depending on the specific job. Anyone who enters the field of communications, however, needs certain basic skills: the ability to express ideas clearly and a solid understanding of the subject to be communicated.

A successful career as a writer, photographer, or filmmaker does not require a college degree in earth sciences. However, those who want to work on science-related issues generally take various science courses as part of their education, often studying more than one area of science. Courses in creative and nonfiction writing, print and broadcast journalism, film and television production, and computer science are all helpful foundations for a career in science-related communications. Majoring in one of these areas is sensible because they all offer specific skills often used in the communications fields generally. A science reporter on television, for example, should have experience with cameras and microphones.

Employment Opportunities. The growth of cable television and expansion of new media such as CD-ROM and the Internet has increased the demand for science programming. A strong market also still exists in traditional print media such as magazines and newspapers. As a result, there are increasing opportunities for work as a science writer, photographer, filmmaker, or television producer.

National television networks, magazines, and newspapers, as well as local media outlets, offer various employment possibilities. Some national magazines, such as *Science Week* and *Discover,* and television programs, such as *National Geographic Explorer* and *Nova,* focus mostly or entirely on science issues. Many other magazines and newspapers have science sections. Professional science associations that produce newsletters are another source of employment.

Some universities, research institutes, corporations, and government agencies hire science writers and producers to create articles and other materials. The Internet provides opportunities for people with computer skills and science backgrounds to create science-related Web sites. CD-ROM publishing is another growing field that provides science-related employment opportunities.

A career in communications can be exciting. Writers, reporters, and photographers often travel to exotic locations for their stories, sometimes remaining away from home for long periods. Work in these careers is not always glamorous, however. Individuals must spend much of their time researching topics at libraries, searching the Internet or other sources of information, and interviewing scientists on the phone or in person.

The National Association of Science Writers, the Association of Earth Science Editors, the American Association for the Advancement of Science are good resources for anyone interested in pursuing a career in communications in the field of earth sciences.

Carson, Rachel
1907–1964
American biologist and environmentalist

Rachel Carson was a biologist who became a pioneer in the American environmental movement. Her writings awakened a new interest in marine life and alerted readers to the dangers of chemical pollution.

Studying the Sea. As Rachel Carson grew up in Springdale, Pennsylvania, her mother encouraged her love of nature and wildlife. Carson showed a talent for writing and studied English in college. She planned to become a writer, but a biology course renewed her interest in science.

During the early 1930s, Carson earned a master's degree at Johns Hopkins University in Baltimore, Maryland. Her special interest in the life of the sea led her to further studies at the Woods Hole Marine Biological Laboratory in Massachusetts. In 1936 she took a position as an aquatic biologist with the U.S. Bureau of Fisheries, which later became the U.S. Fish and Wildlife Service.

Carson had not forgotten her plans to become a writer. During her years with the Fish and Wildlife Service she wrote many leaflets about the bureau's mission—to preserve the nation's wildlife. In 1941 she published *Under the Sea-Wind,* a book about life on the seashore and the ocean bottom. It was so well received that in 1951, after much research, Carson published *The Sea Around Us,* which introduced readers to the history of the Earth and the oceans. This book displayed Carson's gift for writing about complex scientific subjects in beautiful, easy-to-read language, and it became a best-seller.

Environmental Alarm.   In 1952 Carson retired from the Fish and Wildlife Service to devote herself to writing. *The Edge of the Sea* (1955), a guide to the seashore, reflected Carson's growing interest in ecology, the study of the interrelationships among living things and their environment.

In 1962 Carson published her best-known book, *Silent Spring.* It opens with an account of a fictional American town, complete with farms and thriving wildlife. Then a blight arrives, bringing disease and death to animals and plants. The cause of the blight, Carson explained, is the use of chemical pesticides which gradually poison the soil and water. Drawing on published scientific studies, she documented for the ordinary reader some effects of chemical pollution on the environment. She argued that the public should carefully consider the use of these chemicals in light of the potential harm.

Silent Spring caused a sensation. The chemical industry attacked Carson's scientific ability and her work, but the scientific community and public policy makers soon agreed that her warning was appropriate. By introducing readers to ecological issues that would affect their lives, Rachel Carson's work helped spark a growing environmental awareness among Americans. The first Earth Day, April 22, 1970, was one outcome of this new awareness, and a second outcome was the creation of the Environmental Protection Agency (EPA) the same year.

Rachel Carson.

Caves and Karst Topography

Some prehistoric people were living in caves by about 500,000 years ago. Deep underground, where sunlight never enters, such people later painted the rock walls with scenes that illustrated their lives and beliefs. A few cave dwellers still live in parts of Africa, Asia, Europe, and the Near East. The cold, moist underground rooms and tunnels of caves are parts of natural drainage systems. They give researchers and cave explorers (called spelunkers) a window into the world below the Earth's surface. Caves are found in many parts of the world, but they generally occur in a particular kind of topography, or landscape, called karst.

Karst Landscapes.   Most landscapes, such as valleys, are formed by a combination of chemical and mechanical processes. In chemical attack, water (often containing acids and other substances) breaks down minerals in the underlying **bedrock** and forms soils. In mechanical **erosion,** flowing water helps to break down rock and carries away soil and weathered debris. Together these processes sculpture the landscape. Karst landscapes form in regions where chemical attack is the most important part of the process.

bedrock   solid rock that lies beneath soil and other surface cover

erosion   wearing away of land by wind and water

57

Caves and Karst Topography

Interior of the Cave of the Winds, Colorado Springs.

In karst landscapes, streams do not always flow through valley bottoms. Instead they may disappear into holes in the ground or into a pile of loose rocks, leaving only a dry streambed beyond. Such streams are called sinking streams. Karst regions also contain round depressions called sinkholes. Some are small and only a few feet deep. Others are as big as valleys.

Sinking streams and sinkholes collect water and drain it into underground channels. Some of these channels are filled with water. Some are dry or partly dry but have no openings through which people can enter. Channels that do have such openings are called caves.

How Caves Form.   Karst landscapes and caves form where the Earth's surface consists of certain kinds of bedrock—primarily limestone, dolomite, and gypsum. Plain water dissolves gypsum; limestone and dolomite will dissolve in slightly acidic water (water that contains small amounts of acid). The most important acid is carbonic acid, which is formed when carbon dioxide from the atmosphere, from plant roots, and from decaying matter in the soil dissolves in rainwater and streams.

As rainwater moves down through the soil to bedrock and as surface streams flow across rock, water seeps down into cracks and seams, gradually dissolving the rock and enlarging the seams into channels. Eventually the entire stream sinks into the channel, abandoning the surface streambed. Rainwater that gathers in depressions forms sinkholes through a similar process—seeping through surface pores and cracks to dissolve the rock, forming hollow areas beneath.

The underground channels collect water from sinking streams and sinkholes and deliver it to karst springs, where it reaches the surface again. Over time a new layer of seams and tunnels forms at a deeper level, and the stream sinks again. The first layer of underground channels becomes air-filled caves. Erosion, collapsing sinkholes, or other processes create holes through which animals and people can enter.

The Longest and Deepest Caves

Of the tens of thousands of caves that cave explorers have mapped, most are fairly small. A few, however, are truly enormous. The world's longest known cave system, Mammoth Cave in Kentucky, has more than 553 km of interconnected chambers and passages. Caves in mountainous regions are often very deep, descending through steeply sloping passages, vertical shafts, and underground waterfalls. For example, the Huautla Plateau in southern Mexico has cave systems that reach depths of more than 1,200 m.

sediment soils, rock particles, and other materials that are deposited over time and make up the ground, whether on dry land or at the bottom of a body of water

Cave Features.   Some caves consist of single tunnels with few side passages. Some have branching patterns, similar to those of surface streams, while others are mazes that resemble the patterns of city streets and are not entirely connected. In some caves, layers of passages at different levels are linked internally by collapsed roofs or by pits and shafts.

Water that has become very acidic while seeping through the soil dissolves significant amounts of limestone or dolomite when it reaches the bedrock. It carries that dissolved rock with it into the cave, where it may drip from the ceiling or trickle along a wall. In the cave, the water may release carbon dioxide (becoming less acidic) and redeposit its load of minerals. Over time these deposits build up into various forms called speleothems. Speleothems include stalactites, which hang from the ceiling like icicles, and stalagmites, which rise from the floor.

Mineral formation in caves depends upon the chemistry of the seeping waters. Scientists have found more than 270 minerals—most of them quite rare—in caves. Because the minerals form in a constant environment, they often appear as large, perfect crystals of unusual shape. In some caves, such as Jewel Cave in South Dakota, masses of crystals jut out of the walls, looking like fistfuls of gems.

Scientists use various methods to date cave formations. They can measure radioactivity from minute amounts of uranium in speleothems to determine their age. Using this method, scientists have traced speleothem deposition back 350,000 years. A technique called paleomagnetic dating compares the positions of tiny particles of magnetic iron in cave **sediments** with known patterns of magnetism in the Earth's history. Using this technique, scientists have determined that the uppermost, or oldest, passages in the Mammoth Cave system formed 2 to 3 million years ago, at the beginning of the ice ages.

Life Underground.   Living organisms are found in almost every possible habitat on Earth, including caves. Conditions such as darkness and a sparse food supply make caves rather harsh and inhospitable. On the other hand, except for the floods that can occur where streams run through them, caves offer a constant environment to which organisms may adapt.

The Li River and Karst mountains in the Guilin area of southern China.

Bears, birds, and other mammals sometimes use caves as shelters, generally staying near the entrances. Other animals spend a larger part of their life cycles in caves. For example, bats sleep, give birth, and hibernate in caves, although they leave the caves to find food. Still other organisms, including some types of fish, crayfish, salamanders, and beetles, have adapted completely and spend their entire lives in caves. Because they live in permanent darkness, many of these species have lost their eyes and their coloration while developing other means of sensing their surroundings.

In terms of their **ecology,** communities of cave organisms are much less complicated than surface communities. For this reason, caves are useful as ecological and **evolutionary** laboratories where investigators can study the relationships among organisms, environment, and food supply. SEE ALSO WEATHERING AND EROSION.

ecology   science that deals with the inter-relationships among living things and their environment

evolutionary   related to evolution (changes in groups of related organisms occurring over time)

Cenozoic Era

See Geologic Time.

Ceramic Materials

Ceramic materials are naturally occurring, nonmetallic minerals and rocks that can be made into ceramic ware. Clay is the oldest and still the most important ceramic material. Using heat and other processes, artists, craftspeople, manufacturers, and engineers transform clay and other ceramic materials into a tremendous variety of objects.

Clays.   In early times, the only ceramic raw materials were natural clays that could be formed into shapes and dried. The dried pieces were then fired—that is, carefully heated to make them hard and waterproof. Pottery made from fired clays has been traced back to the beginnings of human civilizations. Such objects are extremely important to archaeology, the study of past cultures. The interpretation of ceramic **artifacts** can reveal much about human life in the past.

artifact   tool, artwork, or other object made by humans

Around the beginning of the Christian era, Chinese artisans began making pottery from a weathered and partially decomposed granite clay that contained kaolinite, feldspar, and quartz. This material, together with the development of new, hotter kilns, allowed the Chinese to make watertight pottery and beautiful, smooth ceramic ware called porcelain. The region in central China where this special clay is found is still renowned for the production of high-quality ceramic products.

In the late 1200s, the Italian explorer Marco Polo brought back to Europe samples of the Chinese weathered-granite clay and the porcelain made from it. When a similar type of clay was discovered in Cornwall, a region of southwestern England, it became known as "china clay." The term is still used, especially in Europe, for kaolins—clays that are made up primarily of the mineral kaolinite and are white when fired. Kaolins are found in many parts of the world. Major deposits are in the southeastern United States, southwestern England (Cornwall area), southern and eastern Germany, Japan, China, Indonesia, New Zealand, Australia, Brazil, and Argentina.

organic   related to carbon compounds, which are found in living things; living or formerly living

Another kind of clay, ball clay, contains **organic** matter that gives it a gray or black color. Ball clays are white or near white in color when fired. Underclays (also called seat earths) are strong and easily shaped when mixed with water. Depending on their iron content, some underclays are off-white or red after being fired. Underclays are used for pottery, stoneware, and heavy clay products such as sewer pipe and floor tile. Shales, which generally

STS-37 crew members inspect heat shield tiles on the underside of *Atlantis,* Orbiter Vehicle 104, after landing at Edwards Air Force Base in California.

turn red after firing, are found in large quantities on every continent and are used primarily in heavy products such as floor tile. Several other kinds of naturally occurring clays are also used in ceramics manufacture.

Other Ceramic Raw Materials.   Silica, often in the form of finely ground quartz sand, is mixed with other ceramic materials such as kaolin and feldspar to make porcelain and other ceramic wares. Feldspar is also used in glazes for ceramics. A granitelike rock called nepheline syenite, mined in Canada and Norway, replaces feldspar in some ceramics and glazes. Nepheline syenite is used in the making of sanitary ware, such as sinks and toilets, as well as wall tiles.

The ceramics used for electrical insulators, wall tiles, and catalytic convertors in automobiles include significant amounts of talc. This mineral gives ceramics high thermal shock resistance, which means that they do not crack easily when exposed to extreme heat. Like talc, some other ceramic minerals, including pyrophyllite, magnesite, bauxite, and graphite, are especially useful because they are refractory—they can withstand very high temperatures. These ceramic minerals are used in furnace linings, containers for holding melted steel, and similar applications.

Ceramic Processing.   Whether in an artist's studio or in a manufacturing plant, making ceramics involves several steps. The first, and one of the most important, is to select the proper ceramic raw materials to achieve the intended result.

After the raw materials have been mixed, the mass of material, known as the ceramic body, is formed into the desired shape. There are a number of methods for shaping ceramic bodies, including pressing them by hand or machine, turning them on a potter's wheel (an ancient invention), and squeezing them through mechanical forms. The shaped forms must next be dried under the right conditions. Although makers might prefer, for efficiency's sake, to shorten the drying time, ceramic ware that is dried too quickly may shrink unevenly when fired, causing it to crack.

Once dried, the ceramic pieces are fired in a furnace called a kiln. Firing is the most critical step in the process, since ceramics can be ruined by improper firing. Ceramists and ceramics engineers must know as much as possible about the chemical reactions that take place at various temperature levels. All ceramic ware shrinks during firing; the challenge is to achieve exactly the right amount of shrinkage and the desired texture in the final piece. The firing process is a science in itself, and such details as the placement of pieces within the kiln and the rate of heat increase can significantly affect the outcome of the firing. SEE ALSO REFRACTORY MATERIALS.

Chlorofluorocarbons

See Pollution of Atmosphere.

Chromium

alloy substance composed of two or more metals, or of a metal combined with a nonmetallic substance

A hard, corrosion-resistant, silvery-white metal, chromium is an essential industrial resource that is combined with other metals to form various **alloys.** The primary use of chromium is in the production of stainless steel, made by combining chromium with iron. Stainless steel is one of the most important alloys because of its hardness, strength, heat resistance, and resistance to tarnishing. Stainless steel is used to make household utensils, food containers, automobile parts, weapons, and cutting tools. An alloy of nickel and chromium is used to make heating elements for electric appliances. As the mineral chromite, chromium is used in high-temperature furnace linings. Chromium can also be combined with certain metallic elements to form various chemical compounds that are used as pigments in paints and dyes and for processing hides into leather.

extract to separate a substance from its source

Chromium is a comparatively scarce metal and is **extracted** commercially only from the mineral chromite. Chromite is composed primarily of iron, chromium, and oxygen, with varying amounts of magnesium and aluminum. More than half of all chromite comes from South Africa and from Kazakhstan in central Asia, the site of the world's largest chromite mine. Other countries that produce important amounts of chromite are Turkey, the Philippines, Zimbabwe, Albania, India, and Finland. SEE ALSO ALLOY METALS; INDUSTRIAL MINERALS.

Climate Changes

Although global and regional climates basically remain the same year after year, there have been natural changes of varying magnitudes that have occurred throughout Earth's history. These changes range from minor regional fluctuations to major global shifts, such as those that occurred during the ice ages. Most major changes occur very gradually over the course of hundreds of thousands of years.

Earth scientists have been studying the potential for future climate changes. They are especially interested in the effect that human activities may have on climate. There is concern that these activities have the potential to change Earth's climates much more rapidly than would natural processes. Changes of this sort could have a dramatic effect on all life on Earth.

Role of Natural Processes in Climate Change. There are a number of possible natural causes for the changes in climate that have occurred during Earth's 4.55 billion–year history. The energy output of the Sun has varied slightly over time. Solar energy, in the form of light and heat, is a major factor affecting climate. Changes in the amount of solar energy reaching Earth would thus contribute to climate change, resulting in a cooling or warming of the planet.

Natural climate changes may also occur as a result of slight, predictable changes in Earth's orbit around the Sun and in the tilt of its axis. Scientists know that the tilt of the planet's axis varies over the course of thousands of years. This "wobble" changes the position of continents in relation to the Sun's radiation, sometimes resulting in climatic change.

Another natural process that causes climate change is **continental drift.** Changes in the position of Earth's continents within the oceans expose continents to the various climates of different latitudes and affect the global climate by changing the amount of ice, distribution of mountains, and paths and temperatures of ocean currents. The build up and erosion of mountains over time can also produce changes in temperature, rainfall, and winds over large regions.

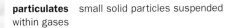

continental drift theory that the continents move over the surface of the Earth on large segments of Earth's crust and uppermost mantle (region just below the crust)

Gases and **particulates** in the atmosphere can reduce the solar energy that reaches Earth's surface. Throughout the history of the planet, volcanic activity has released large amounts of ash and gas into the atmosphere, temporarily lowering temperatures worldwide. For example, the eruption of Mount Pinatubo in the Philippines in 1991 resulted in a slightly cooler climate over much of the globe for a period of a few years.

particulates small solid particles suspended within gases

Human Impact on Climate. Over the past few hundred years, human activities have caused changes in Earth's atmosphere that many scientists believe may contribute to climate change. Humans may have the greatest impact on climate through the burning of **fossil fuels,** which releases carbon dioxide (CO_2) and other gases into the atmosphere.

The atmospheric concentrations of so called "greenhouse gases" are increasing. These gases—most notably CO_2, methane, chlorofluorocarbons, and nitrous oxide—trap heat within Earth's atmosphere, preventing it from escaping into space. This is known as the "greenhouse effect."

Some concentration of greenhouse gases is necessary to maintain Earth's present climate. Many scientists, however, are concerned that the rapid increase in these gases could lead to a substantial **global warming** in a relatively short time. This warming could drastically alter not only global temperatures but also patterns of precipitation, wind patterns, and ocean currents and levels. Such changes could severely disrupt **ecosystems** and endanger forms of life that could not adapt quickly to new climate and environmental conditions.

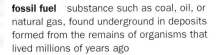

fossil fuel substance such as coal, oil, or natural gas, found underground in deposits formed from the remains of organisms that lived millions of years ago

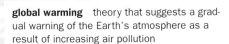

global warming theory that suggests a gradual warning of the Earth's atmosphere as a result of increasing air pollution

ecosystem system formed by the interaction of a group of organisms and their physical environment

Some greenhouse gases are released into the atmosphere through natural processes. Forest fires release CO_2, and methane is produced by the natural decay of plant and animal matter. However, human activities are producing increasing amounts of these gases. The burning of fossil fuels releases huge quantities of CO_2, and **deforestation** removes vegetation that absorbs CO_2 from the atmosphere. Methane is produced during coal mining and petroleum production and by the burning of **biomass fuels.** Chlorofluorocarbons, chemicals used in spray cans as propellants and in refrigeration and air conditioning, are also released during various human activities. Although the use of chlorofluorocarbons has decreased over the past decade, their concentration in the atmosphere continues to increase.

deforestation clearing trees from an area without replacing them

biomass fuel renewable fuel resource derived from living and nonliving plant matter

Human activities have clearly altered the chemical composition of the atmosphere. Scientists are still unsure exactly how this change will affect the climate in the long run.

Conflicting Viewpoints

Although evidence suggests that global warming is occurring, there is disagreement among scientists over the causes and impact of this change. Some scientists argue that recent increases in global temperature are part of natural fluctuations in climate, and that human activities have had minimal impact. It is difficult to draw any conclusions at this point because the temperature increase is still within the range of natural climate variability. Furthermore, it may be years before the real causes and impact of current global warming will be known for certain.

Climatology

oceanography science that deals with oceans

meteorology science that deals with the atmosphere, especially with regard to climate and weather

biosphere all of Earth's regions that support life; all of Earth's living organisms

cryosphere frozen water in all its forms on the surface of the Earth

photosynthesis process by which plants manufacture food from sunlight

ecosystem system formed by the interaction of a group of organisms and their physical environment

Climate Modeling. Scientists try to determine what will happen to the climate by creating computer models based on mathematical equations. Taking into account what is known about winds, ocean currents, and other variables, these models can make predictions about the possible effects of increasing levels of greenhouse gases (primarily CO_2) on temperature, precipitation, and other aspects of climate.

Accurate predictions are difficult to make, however, because climate is so complicated and because there are so many different variables. Still, climate modeling improved dramatically during the 1990s, and scientists are increasingly in agreement about what climate changes may take place over the next 100 years.

Many scientists agree that global warming will continue unless people make drastic changes to reduce the amounts of greenhouse gases released into the atmosphere by human activities. Among the potential effects are a rise in sea levels, higher temperatures in most regions, significant changes in precipitation, and more unpredictable and extreme weather conditions. Such changes would have a major impact on all life around the globe. For this reason, climate change may be one of the most important issues of the coming decades. SEE ALSO ATMOSPHERE, EARTH'S; CLIMATOLOGY; CONTINENTS, EVOLUTION OF; EARTH, MOTIONS OF; EARTH AS A DYNAMIC SYSTEM; ENVIRONMENTAL CHANGES; FOSSILIZATION AND THE FOSSIL RECORD; GEOLOGIC TIME; GLACIAL AGES; PLATE TECTONICS; POLLUTION OF THE ATMOSPHERE.

Climatology is the study of climate, the factors that determine climate, and the changes that occur in climate over time. The study of climatology involves several branches of science, including chemistry, biology, geography, geology, **oceanography**, and atmospheric science, or **meteorology**.

Weather and Climate. Weather is the state of the Earth's atmosphere at a particular place at any given moment. Characterized by short-term changes in atmospheric conditions, it includes such factors as storm and cloud movement, the likelihood of precipitation, and changes in humidity.

Unlike weather, climate consists of the expected or general weather conditions over periods of years or longer. Although it is determined by the same variables as weather—temperature, precipitation, winds, and humidity—a region's climate is more than an average of these variables. It includes the extremes in the weather as well as the general range of expected weather. Record high and low temperatures, record-breaking floods and storms, and similar events are all parts of the climate.

Components of the Climate System. The Earth's climate system consists of five components: the atmosphere, **biosphere**, lithosphere, hydrosphere, and **cryosphere**. The primary component is the atmosphere, an area of gases, dust particles, and water vapor that extends above the Earth's surface into space. The atmosphere sustains life by providing oxygen for animals to breathe and carbon dioxide for plants to use in **photosynthesis**; it also shields living organisms from hazardous radiation from the Sun. The atmosphere is also the area in which weather conditions vary on a daily basis.

A second component of the climate system is the biosphere—the total of all Earth regions that support life. The biosphere is divided into many varied **ecosystems,** each with distinct climates and groups of living organisms. The biosphere affects and is affected by climate. Changes in climate can upset the delicate balance among the living and nonliving components of many ecosystems.

erosion   wearing away of land by wind and water

groundwater   underground water that supplies natural springs and can be tapped by wells

precipitation   water that falls to Earth in the form of rain, sleet, or snow

permafrost   permanently frozen ground found in arctic regions

latitude   distance north and south of the equator, measured in degrees

The lithosphere is another component of the climate system. It is the rocky outer layer of the Earth, including the crust and the uppermost portion of the mantle (just below the crust). The lithosphere changes continually as a result of **erosion,** volcanic eruptions, earthquake activity, and other geologic processes. These changes affect climate over long periods of time.

A fourth component is the hydrosphere—oceans, lakes, rivers, **groundwater,** and all other water on Earth. The hydrosphere plays an important role in climate. Ocean currents carry warm and cold water to different areas on the planet, affecting both weather and climate. Through the process of evaporation, the hydrosphere is the source of all **precipitation.** Changes in sea levels, the position of ocean currents, and other aspects of the hydrosphere over long periods of time can have profound effects on climate.

The final component of the Earth's climate system is the cryosphere— frozen water in all its forms. The cryosphere includes glaciers, snow cover, frozen lakes and rivers, and **permafrost.** Except for seasonal snow and ice in many areas, most of the cryosphere is located in the polar regions and on mountains where the temperature is low enough to permit ice to remain throughout the year.

The cryosphere is very sensitive to changes in temperature, and its extent is affected by seasonal and long-term variations in climate. During the ice ages, glaciers in many parts of the world expanded dramatically, contributing to cooler climates around the world. A retreat or melting of glaciers, on the other hand, contributes to warmer climates. The ice sheets and cold temperatures of the polar regions have a profound effect on climate systems. They contribute to weather and climate by means of the exchange of energy between the polar and tropical regions via air and water currents in the atmosphere and oceans.

Factors That Determine Climate.   Climate is determined by a variety of factors, or climate controls, including **latitude,** air circulation, ocean currents, and the local geographic features of an area.

The most important factor in climate is latitude, because it influences the amount and intensity of solar radiation that reaches the surface of the Earth. Solar radiation is energy that comes from the Sun in the form of light and heat. This energy is distributed unequally over the Earth due to the rotation of the Earth on its axis, the tilt of the axis, and the planet's revolution around the Sun.

The Sun's rays hit the Earth directly at the equator and almost directly near the equator, concentrating the energy over a small surface area. This concentrated energy causes warm climates throughout the year. At higher latitudes (areas closer to the poles), the Sun's rays hit the Earth more indirectly, causing the energy to spread over larger areas and resulting in generally cooler climates. Seasonal changes in both hemispheres are caused by the tilt of the Earth's axis in relation to the planet's revolution around the Sun. In each hemisphere, the Sun's rays strike the Earth at the higher latitudes more directly during the summer than during the winter.

The unequal distribution of solar energy across the surface of the Earth creates the forces that drive the climate system, causing pressure imbalances that affect air circulation and ocean currents. This process operates as if the Earth were a large heat engine; it balances the unequal heating at the surface by redistributing heat from the tropical regions to the poles. This redistribution of energy is carried out through movements of air and water.

Air circulation patterns are an important factor affecting climate. Close to the equator, air near the Earth's surface is heated by the Sun and becomes

Climatology

Time series of the trend in atmospheric carbon dioxide since 1800 as inferred from ice core analysis and by direct measurements.

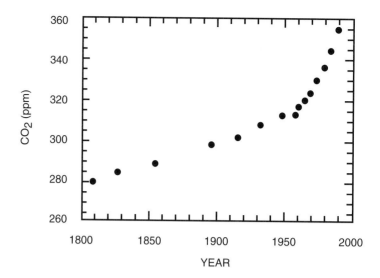

Trend in global surface air temperatures (shown as a difference from the 1960–1990 global average) taken from measurements.

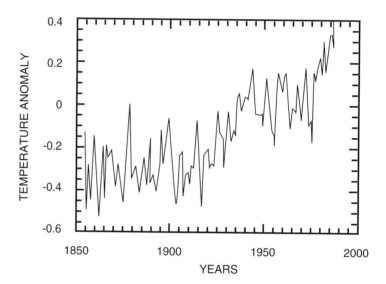

less dense, which causes it to rise. This process creates low-pressure centers—large air masses in which the air pressure is lower than in surrounding areas. These centers carry storms and are the means by which the atmosphere mixes warm air and cooler air from the higher latitudes. The overall movement of air creates global patterns of atmospheric circulation. The winds that are part of this circulation pattern carry both heat and moisture from one place to another.

Ocean currents, caused primarily by winds and the Earth's rotation, help redistribute warm and cold masses of water from one area to another. These currents influence the climate of any nearby regions. For example, the Gulf Stream is an ocean current that carries warm water from the Caribbean Sea to the western part Europe, helping to warm the climate of that region. Because large bodies of water change temperature slowly, oceans reduce climate extremes in the land areas they influence. Coastal areas, for example, generally have warmer winters and cooler summers than inland areas.

Local geography also affects climate in a number of ways. Mountain ranges, for example, influence the distribution of precipitation by acting as barriers to large air masses. Throughout the world, more rain typically falls on the windward side of mountains. Altitude is another factor in climate. Because temperatures decrease with increases in altitude, places of high elevation are typically cooler than areas of low elevation.

Types of Climate. There are many distinct climates on Earth, each characterized by variations in temperature, rainfall, and types of vegetation. The world's climates can be organized into four major regions: tropical, midlatitude, high latitude, and dry.

Tropical climates are found in the tropics—areas distant from the equator to about 30° latitude (north and south). Their basic feature is warm temperatures that change little during the year. There are two main types of tropical climate. The tropical rain forest climate is wet throughout most of the year. The combination of warmth and moisture produces dense rain forests. The tropical savanna climate has very distinct dry and wet seasons that produce extensive grasslands.

Midlatitude climates are found between the tropics and about 60° latitude (north and south). These climates are noted for their great variety, their changing seasons, and the wide range of vegetation types they support. There are four main types of midlatitude climates. Found in coastal areas, a marine west coast climate has mild, rainy winters and cool summers. Another coastal climate, the Mediterranean climate, has mild, rainy winters and hot, dry summers. Inland areas away from oceans often have a humid continental climate, characterized by long, cold, snowy winters and short, very hot summers. The fourth type of midlatitude climate, the humid subtropical climate, is found near the tropics. It has hot, humid summers with generally heavy rainfall and short, mild winters.

High latitude climates are found between the midlatitudes and the poles. Their basic feature is cold temperatures throughout the year. One type of high latitude climate, the subarctic climate, lies just below the Arctic Circle. It features very cold winters and cool summers that support huge evergreen forests in some areas. The tundra climate is cold and dry, with too little moisture and warmth to support the growth of trees. The polar climate, found near the poles, is too cold for almost all types of vegetation. Although it may be found anywhere on Earth, a highland climate is considered a high latitude climate because it is characterized by cool or cold temperatures year-round. This type of climate is found in mountainous areas at high elevations.

Dry climates are also found in various places on Earth. Their basic feature is little or no rainfall for long periods of time. Temperatures can range from very hot to very cold. There are two basic types of dry climate. Desert climates are the driest climates on Earth, with less than 25 cm of rainfall a year. With so little water, few plants can survive in a desert climate. A steppe climate is one that averages 25 to 50 cm of rain a year. This amount of rainfall is enough to allow the growth of bushes and short grasses, which often cover the landscape. SEE ALSO ATMOSPHERE, EARTH'S; BIOSPHERE; CLIMATE CHANGES; EARTH, MOTIONS OF; EL NIÑO AND LA NIÑA; GLACIAL AGES; GLACIERS AND FROZEN WATERS; HYDROSPHERE; LITHOSPHERE; MANTLE; PERMANENTLY FROZEN GROUND; PRECIPITATION; SOLAR ENERGY; WIND.

Extremes and Records

Record temperatures and rainfalls reflect the great extremes of the Earth's climate system. The hottest temperature ever recorded was 58°C (136°F) at El Azizia, Libya, on September 13, 1922. The lowest recorded temperature was –89° (–129°F) degrees at the Vostok research station in Antarctica on July 21, 1983. The record rainfall for one day—183 cm—occurred on the island of La Réunion in the Indian Ocean on January 7–8, 1966. The record for annual rainfall is 2,647 cm, which fell in Cherrapunji, India, between August 1860 and August 1861. At the other extreme, parts of the Atacama Desert in Chile received no rainfall at all for a period of 14 years.

Coal

Coal is an **organic** type of **sedimentary rock,** formed from plant matter that accumulates as **peat** in waterlogged areas such as swamps. Because it is abundant and combustible, coal is widely used throughout the world as a source of energy.

organic related to carbon compounds, which are found in living things; living or formerly living

sedimentary rock rock formed from deposits of sediments (soils, rock particles, and other materials) over long periods of time

peat dark brown or black residue of plant material that has partially decomposed (rotted) underwater

sediment soils, rock particles, and other materials that are deposited over time and make up the ground, whether on dry land or at the bottom of a body of water

Formation of Coal

Coal is formed by a sedimentary process in which layers of organic matter that have accumulated over many thousands of years are deeply buried under newer deposits of organic matter and other **sediment.** The manner and length of time in which this occurs determines how much of the peat in a particular deposit turns into usable coal.

Coalification.   In its natural state, peat is about 70 to 90 percent water. As additional layers of peat, mud, silt, and sand accumulate, the peat is compacted so that the water is squeezed out. If enough layers accumulate, peat is eventually transformed into rocklike coal from the pressure of the layers on top (called the overburden) and the increasing temperature of the depths at which it is buried. This process is called coalification.

When peat is near the surface, the plant debris is attacked by chemical agents, such as oxygen and enzymes, and biological agents, such as bacteria and fungi. Less stable parts of the debris, such as cellulose and proteins, decompose; more stable parts, such as resins and waxes, tend to remain unchanged. As the peat is buried more deeply, however, pressure and temperature rise and eventually transform these stable components into even more stable substances that make up coal.

Rank and Grade of Coal.   The degree of coalification, commonly called the "rank" of coal, depends upon several factors. These include the deepest point at which the peat has been buried, the resulting pressure and temperature created by the overburden, and how long the peat was exposed to these conditions. Generally, the coalification process takes millions to tens of millions of years; higher ranks of coal are formed by higher pressures and temperatures and over longer periods of time. The lowest rank of coal is called lignite, followed in order by subbituminous coal, bituminous coal, and finally anthracite, which is the highest rank. The higher the rank of the coal, the higher its carbon content.

Coal Canyon in Arizona.

Higher ranks of coal can generate more heat when burned. However, the combustibility of coal is also affected by its purity, or grade. When peat is formed, minerals may be carried into it by wind or water. These impurities lower the grade, and reduce the combustibility, of the coal that is eventually produced. Unless at least 50 percent of a rock formed from plant debris is combustible, it does not qualify as coal.

Types and Components of Coal

Depending on the environment in which it forms, peat can be made up of plant material of many kinds that breaks down to varying degrees. These variations produce several types of coal, many of which can be distinguished with the naked eye or through a hand lens. One major distinction is between banded and nonbanded coals.

Coal Lithotypes.   Banded coals are the most common types. Their name comes from the fact that one can see separate layers, or bands, of material in the coal. These layers, called lithotypes, are created by the varied kinds of plant matter from which the coal was formed. They are particularly easy to see in bituminous coal, which contains shiny layers called vitrain, semidull layers of clarain, and dull layers of durain. Vitrain is composed mainly of larger pieces of wood and bark from stems, branches, and roots. Clarain and durain, which together are referred to as attrital coal, are composed primarily of finer fragments of plant material. Dull attrital coal usually has a rough, grainy texture and is very hard. It tends to contain a higher percentage of fine mineral matter, particularly clay and quartz.

Another lithotype found in banded coal is fusain, a crumbly, charcoal-like material that usually occurs throughout a coal bed. It may form layers of its own or occur in the other lithotypes. Fusain has large pores that are often filled with mineral matter and it tends to be concentrated in places where blocks of coal separate easily from the coal bed. Fusain is the material in coal that turns the hand black when it is handled.

As their name implies, nonbanded coals have no layers. They appear to be composed of fine materials and have a dull, greasy sheen. This is because they are made up of pulverized plant material that was deposited in still water. Nonbanded coals typically contain pollen and spore grains as well as the remains of algae; they can be identified only with the help of a microscope.

A third type is impure coal, which is 25 to 50 percent mineral matter by weight. The mineral impurities often include fine-grained substances such as clay, silt, and sand that were washed or blown into the peat while it accumulated. Metal sulfides and carbonates can also be found in pores and fractures in impure coals.

Macerals, Minerals, and Pores.   Other components of coal that cannot be seen without a microscope are plant-derived structures called macerals. Despite their small size, some macerals can be identified and classified according to the type of plant structures that formed them. For example, sporinite comes from spores, alginite from algae, and resinite from resins. Other macerals are produced by processes that alter their original structure. Fusinite, for instance, is created by **oxidation**, which can occur during forest fires and other processes.

Noncombustible minerals may also occur in coal, reducing its value as a source of energy. They typically include clays, pyrite, quartz, and calcite.

The Environment Impact of Coal Mining

Coal mining is often criticized because of the harm it can do to local environments. Surface mining techniques often require entire mountaintops to be removed or deep pits to be dug in order to reach coal deposits. These methods destroy not only the mountain but also any vegetation that grows in the mining area. Mining companies typically need to obtain special permission to carry out such activities. From another perspective, however, removal of mountaintops may be beneficial in a particular instance. In the Appalachian region of the United States, mountaintop removal has created valuable flat surfaces for commercial, agricultural, and residential uses.

oxidation combination of a substance with oxygen

Some minerals can be recognized with the naked eye or a hand lens, but many require a microscope to be seen. In the 1960s, a technique called low-temperature ashing was developed to make recognition of minerals easier. In this process, powdered samples of coal are heated in a very oxygen-rich environment at temperatures of 120° to 150°C, burning off the plant-derived material but leaving the minerals unchanged.

The pores throughout the internal structure of coal are so small that although they account for very little of the volume, they greatly increase the internal surface area. In fact, each gram of coal has an internal surface area of several hundred square meters. Liquids and gases do not easily pass through such small pores. Methane, a highly combustible gas, collects on the surfaces of these pores and is released during mining. The presence of methane causes a potential hazard and increases the cost of mining, especially for certain bituminous coals.

Technological Properties of Coal

Coal can also be classified by its technological properties, such as washability, coking, and combustion. These properties help determine the commercial value of the coal.

Washability and Coking. All coal contains noncombustible minerals that remain as ash after the coal is burned. Raw coal may contain as much as 50 percent ash, whereas clean (or washed) coal generally has 5 to 15 percent ash. Engineers can predict how much usable coal can be recovered from a deposit by floating samples in liquids of different **densities** to see how much of the coal sinks or floats. The results can help them determine how to operate a cleaning plant to obtain the maximum amount of usable coal.

Coke is a residue left by certain coals after they have been heated in an airless environment. It is rich in carbon and is used to produce metallurgical coke, important in the **extraction** of iron from iron **ore**. A blend of bituminous coals is used to make metallurgical coke with specific properties.

Combustion. Since most coal is burned to generate steam in electric power and industrial plants, its combustibility is of primary importance. The heating value of a coal depends on both the heat content of the organic material and the amount of noncombustible matter (moisture and minerals) present. Generally, coal of higher rank has greater heat content. Higher ranks of coal also require higher temperatures to ignite than do those of lower ranks. Because higher ranks of coal contain less volatile matter, they also burn more slowly and with a shorter flame.

Mineral and ash content are also important during combustion. Ash formed during the burning of coal can produce undesirable deposits called slag. The higher the temperatures required for ash to form slag, the more desirable the coal.

One drawback of coal burning is that it may release harmful chemicals into the atmosphere, contributing to **acid rain** and other pollution. The more mineral impurities the coal contains, the more serious this problem becomes.

Coal Production, Resources, and Uses

Coal is widely available throughout the world, and most countries have coal deposits that can be mined. In 1913 the International Geological Congress in

density amount of mass (matter that causes an object to have weight) in a unit of volume

extraction removal from the Earth

ore metal-bearing mineral or rock

acid rain rain containing chemical pollutants that can harm lakes and streams, wildlife, forests, crops, and structures made by humans

Toronto compiled the first list of the world's coal resources. Since then, the World Energy Council (WEC) has kept track of coal resources and production around the world. The former Soviet Union, China, and the United States have the greatest resources of coal. They are also the three largest producers, with China leading the list of producing nations. In most cases, the rate of coal production of the 1990s could be maintained for several hundred years without exhausting known reserves.

The generation of steam in electric power plants accounts for the greatest portion of coal burned worldwide. In the United States in 1992, 87 percent of the coal burned was used for this purpose. Other important uses include coke production for iron and steel manufacture in industrial plants. Coal may also be converted into liquid and gas fuels, although this practice is not widespread. SEE ALSO ENERGY USE AROUND THE WORLD; MINERALS; MINING; POLLUTION OF THE ATMOSPHERE; SEDIMENTS AND SEDIMENTARY ROCKS.

Coastal Erosion and Deposition

Erosion and deposition are the most important processes that shape the world's beaches. Beaches are worn away by **erosion** as ocean waves pick up and carry soil, sand, and larger particles of rock out to sea. Beaches are also built up by deposition, in which new soil, sand, and larger rock particles are carried from other places and laid down in beach areas.

erosion   the wearing away of land by wind and water

Types of Beaches

There are many types of beaches around the world, ranging from flat to steep, from white to black, and from sandy to rocky. The appearance of a beach tells a great deal about its composition and formation.

Beach Geography.   Most of the beaches along the eastern seaboard of the United States are very flat, and many of them have low barrier islands lying

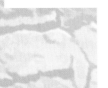
Sandbags and bulldozers are used to protect an eroded beach from further losses.

mantle   region of the Earth between the molten core and the outer crust

continental plate   large segment of Earth's crust and uppermost mantle (region just below the crust) that supports a major landmass

weathering   physical and chemical breakdown of rock that is stationary on Earth's surface

ore   metal-bearing mineral or rock

just offshore. These islands can become submerged during hurricanes and as a result of the rise in sea level caused by glacial melting. As the sea moves up and over the islands, it causes further erosion on the coast. By contrast, most western U.S. beaches are elevated, with steep cliffs running right down to the edge of the sea. The difference between the two coasts involves the division of the Earth's crust and upper **mantle** into large segments, or plates, that slowly move and often collide with one another. Such collisions are lifting the West Coast's **continental plates** up and over bordering oceanic plates faster than erosion can lower the coastline, while much of the East Coast is subsiding, or sinking.

Beach Composition.   Beaches also vary according to the particle sizes and mineral composition of the materials of which they are made. Most beaches consist of sand-sized particles of quartz, feldspar, and calcite—minerals that are usually tan to white in color. These particles are produced by the **weathering** of rocks, such as granite, that are abundant on all continents, and of shells and other parts of sea creatures. These beaches usually also show a few dark grains of heavier minerals, such as garnet and magnetite (an iron **ore**). Sometimes it is possible to trace the sources of these heavier minerals. Some of them come from the erosion of sea cliffs that supply sand to the beach; other minerals may come from rivers that carry particles of rock to the beach from hundreds of kilometers away.

Since heavy minerals wear away more easily than quartz, some beaches are composed almost entirely of quartz sand grains. These beaches are very white in color and can be found at beach resorts around the world. Another type of beach popular with tourists is composed of calcite, a mineral formed from parts of marine animals and plants such as coral, mollusks, and algae. Calcite sand is very white and very soft.

"Black sand" beaches, on the other hand, are formed in places—such as Oregon and California—that are naturally rich in heavy minerals. Such beaches remain dark because waves easily pick up the lighter quartz and feldspar particles and carry them out to sea, leaving behind only the darker, heavier minerals. Black sand beaches often contain valuable elements such as gold, platinum, iron (as magnetite), and titanium. However, not all black sand beaches are made up of heavy minerals. Some consist of particles of volcanic rock that can form when hot lava flows into cold seawater and disintegrates. These types of beaches are found on volcanic islands such as Hawaii.

There are also beaches that have very little sand, being mainly composed of pebbles, stones, and boulders. These types of beaches are quite common in some parts of the world, such as the coasts of England where the flat pebbles and cobblestones are known as shingle. In fact, the term "beach" comes from an ancient Anglo-Saxon word that referred to shingle.

Beach Formation

Both waves and currents move sand and pebbles to produce a beach's shape and slope. The term *morphology* is used to describe the shape and slope of a beach.

Waves and Wave Action.   Most of the energy for beach formation is supplied by waves. Waves produced by storms thousands of kilometers out to sea can help shape a beach. The size and power of such waves depend on the speed

Seawalls: A Mixed Blessing

Seawalls are sometimes built to protect development near a beach, especially in places where it is too difficult or too expensive to bring in new sand. However, building a seawall can produce a number of undesirable effects as well. Seawalls can hinder access to a beach and make it less attractive for recreational purposes. Such reduced access may have a negative impact on businesses that depend upon beach tourism. There is also evidence that seawalls can result in increased erosion of the beach during storms, even when they protect communities successfully.

jetty (*pl.* **jetties**)   structure built out into the sea, a lake, or a river to counteract the effects of tides or currents

of the wind as well as the size and duration of the storm. The longer a storm lasts, the more time it has to transfer energy to the waves.

Although waves lose little of their energy crossing deep ocean waters, they expend most of that energy breaking on the sloping beach when they reach a coastal area. The size and power of the waves determine their ability to move soil, sand, and rock particles from place to place, shaping the beach.

The slope of a beach depends on the energy of the breaking waves and the size of the particles that make up the beach. On pebbly or rocky beaches, breaking waves push the pebbles inland, where they pile up. The water returning to the sea loses most of its energy as it flows down through spaces between the pebbles. The pebbles thus remain heaped up in piles, giving the beach a steeper slope. On a sandy beach, very little water flows down between the fine grains of sand, so the water returning to the sea keeps more of its energy—or power to carry material back with it—creating a flatter beach. The larger the waves, the more material is carried out to sea. This often happens during large storms, such as hurricanes, as well as during the winter, when waves tend to be higher than they are in summer.

Currents.   When waves reach the beach, they generate different types of currents, depending upon the angle at which they break. Waves that break with their crests parallel to the beach can form strong rip currents that flow back out to sea. Rip currents carry sand offshore, eroding the dry portion of a beach and forming bays. Stronger rip currents form larger bays that can cut entirely through the dry part of the beach, allowing waves to reach buildings along the beach.

Waves that break at angles to the beach generally do not form rip currents. Instead they form longshore currents that run along the length of the beach and carry sand down the coastline. Often such currents form beaches by carrying sand from an eroding sea cliff or the mouth of a river and depositing it further down the coast.

Sometimes longshore currents are blocked by **jetties** or other harbor structures. Such structures cause sand to pile up, building the beach out into the sea on the current side of the structure; on the far side, the structure blocks sand deposition. Since waves continue to remove sand from the beach on the far side of the structure, extreme erosion often occurs near jetties and other harbor structures. To prevent the loss of the beach, and the possible destruction of buildings along the beach, sand is often moved into the affected area. It may be simply trucked in from another part of the beach, or it may be pumped in from the seafloor offshore—a process known as beach nourishment. SEE ALSO GLACIERS AND FROZEN WATERS; HURRICANES; MINERALS; PLATE TECTONICS; VOLCANOES; WEATHERING AND EROSION.

Comets

Comets are small bodies of ice and dust that travel around the Sun. Glowing and trailing tails of light, they pass slowly across Earth's night sky for weeks or even months and then disappear until their orbits again carry them close to the inner planets. Many early civilizations thought that comets were warnings of coming disaster or death. Later, these spectacular sky travelers were recognized as cosmic fossils that have changed very little since the earliest days of the solar system, about 4.55 billion years ago.

How Comets Look from Earth.   Comets trail clouds of glowing gas and dust; their name comes from a Greek word meaning "the long-haired one."

Comets

solar wind   flow of charged atomic particles, mostly hydrogen ions (H+), away from the outer layers of the Sun

astronomer   scientist who studies space and the various bodies it contains

When a comet is far from the Sun, it is simply an inactive icy body. As it approaches the Sun, however, heat causes the ice to vaporize, surrounding the comet with a cloud of gas and dust called a coma.

Pressure from sunlight and the **solar wind** pushes the comet away from the Sun, forming two tails. The dust tail is usually curved and appears yellowish because the dust particles scatter sunlight. The gas, or ion, tail is straight and may appear blue because it contains carbon monoxide ions (CO+). As the comet nears the Sun, these tails point behind the comet. When the comet moves away from the Sun, however, its tails lead.

Types of Comets.   British **astronomer** Edmund Halley, in the late 1600s and early 1700s, was the first to use the new theory of gravitation to calculate the orbits of comets. Noting that the comets observed in 1531, 1607, and 1682 had very similar orbits, he decided that all three had been the same comet, and he predicted that it would return in 1758 or 1759. It did, and it was called Halley's Comet in his honor.

Since Halley's time, astronomers have divided comets into two groups. Short-period comets orbit the Sun in less than 200 years, and their appearances are predictable. Halley, which appears every 76 years or so, is one of the brightest of the short-period comets. Long-period comets take longer than 200 years to orbit the Sun. They are often brighter and more active, but their appearances are not predictable.

When Comet Halley swung by Earth in 1986, a small fleet of spacecraft approached it for close study. One of them, the *Giotto* spacecraft, later approached close to Comet P/Grigg-Skjellerup. These missions, together with ground-based observations, have yielded much information about Halley and about comets in general.

Structure of Comets.   The Comet Halley studies confirmed an earlier theory about the makeup of comets. All of the comet's activity comes from a small, solid body called a nucleus. Although the internal structure is uncertain, Comet Halley's nucleus is made up of ices and dust. Of the ices, about

80 percent are water. A hard crust covers most of the surface of Halley's nucleus; the rest is exposed ice. If the crust were to cover all of a comet's nucleus, that comet would stop being active and would look like an asteroid. Comet Wilson-Harrington, discovered in 1949, became inactive and was lost until it was rediscovered as minor planet 1979 VA. This object was later given the asteroid number 4015.

A number of studies have focused on the chemical composition of comets. Observers have detected water, carbon monoxide, carbon dioxide, methane, formaldehyde, and other chemical molecules in comets. Similar proportions of molecules exist in dense molecular cloud cores—the places where stars form. This likeness leads scientists to believe that comets contain the same raw material from which the Sun and planets were formed. Cometary material is of particular interest to scientists because it is the most primitive material in the solar system.

Comets contain a high percentage of both water and **organic** material, in this case molecules made up mainly of carbon, hydrogen, oxygen, and nitrogen. This discovery may link comets to the history of life on Earth. Some scientists believe that at least part of the Earth's water came from comets that crashed into the planet. Comets may also have brought the organic material that was the source of the first life-forms.

Sources of Comets. In 1950 a Dutch astronomer, Jan Oort, claimed that comets originated in a cloud or shell of material located between 20,000 and 100,000 **astronomical units** from the Sun. Such a shell does exist, and it is called the Oort cloud. However, although the Oort cloud is a source of comets, many astronomers do not believe that comets formed there.

Many scientists think that the material in the Oort cloud is left over from the formation of the giant planets Neptune and Uranus. Building blocks from these planets could have scattered across space to form the cloud. From time to time, the gravity of a passing star forces one of these chunks out of orbit, and it falls into the inner solar system to appear as a comet.

Another possible source of comets is the Kuiper Belt, a ring of icy bodies that some astronomers think revolves around the Sun beyond the orbit of Pluto. If the Kuiper Belt does exist, fragments from it may sometimes enter the solar system to join the Oort cloud or to travel in new cometary orbits.

Comet Shoemaker-Levy Strikes Jupiter. In the summer of 1994, people watched for the first time as a comet struck a planet. More telescopes observed this event than any other in history.

The comet in question was Shoemaker-Levy 9, which had split into 21 icy fragments as it approached Jupiter, the solar system's largest planet. Many people compared the comet fragments stretched across space to a string of pearls. Excitement was intense when astronomers who had studied the path of the fragments announced that they would strike Jupiter. Although the impacts took place on the side of the planet that was turned away from the Earth, Jupiter's rapid rotation brought the impact sites into view within a few minutes. NASA's *Galileo* spacecraft had a direct view of the impacts.

Astronomers labeled the fragments and their impact sites from A through V. Fragment A produced one of the largest impact sites, about twice the size of the Earth. Most impacts were surprisingly dramatic, producing visible brightenings of Jupiter's atmosphere.

Scientists studying the observations of the impacts have learned much about the chemical makeup and physical structure of Jupiter's upper atmosphere. At the same time, the collision of Comet Shoemaker-Levy 9 with

organic related to carbon compounds, which are found in living things; living or formerly living

astronomical unit (AU) Earth's distance from the Sun, or approximately 150 million km

Comets of the 1990s

Three remarkable comets brightened Earth's night skies during the 1990s. In addition to Shoemaker-Levy 9 and its collision with the planet Jupiter, that decade produced Hyakutake and Hale-Bopp. Hyakutake passed within 15 million kilometers of Earth and was visible to the naked eye for three months in the spring of 1996. Amateur astronomers Alan Hale and Thomas Bopp discovered Comet Hale-Bopp, an extremely bright comet, when it was beyond the orbit of Jupiter, making it the most distant comet ever discovered by amateurs. The comet was clearly visible, without magnification, to observers on Earth for weeks in the spring of 1997.

NASA National Aeronautics and Space Administration, the U.S. space agency

Jupiter also made many people wonder what would happen if a comet were to strike the Earth. Eugene Shoemaker, one of the comet's discoverers, estimated that an impact with fragment G, the largest, would release several hundred times as much energy as all the nuclear weapons in the world.

Comets and the Earth. The orbits of many comets cross that of the Earth. It is not impossible that a collision could occur. At the edge of Mexico's Yucatán Peninsula is a crater 100 km across, created about 66 million years ago when a large object—a comet or an asteroid—crashed into the Earth from space.

Other such impacts have taken place from time to time in Earth's history. Scientists are studying the possibility that worldwide climate changes caused by large impacts may have led to mass **extinctions**, including the disappearance of the dinosaurs. Some scientific and government agencies are making plans to predict and possibly prevent future impacts.

Yet NEOs (near-Earth objects) such as comets present possible resources as well as dangers. The same comets that may have "seeded" the Earth with water and organic chemicals to enable life to arise may one day become sources of raw materials for space mining operations. It would be far less costly to extract raw materials from comets than to carry them into space through Earth's **gravitational field**. Perhaps comets will supply water, fuel, and other needs for human activities in space. SEE ALSO ASTEROIDS; EXTINCTIONS; IMPACT CRATERING; SHOEMAKER, EUGENE MARK; SOLAR SYSTEM.

extinction dying out completely, leaving no direct descendants (refers to a specific grouping of organisms)

gravitational field area influenced by gravity

Construction Materials

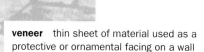

Rocks and mineral products have been used in construction since before the beginning of recorded history. Nonmetallic minerals and rocks are still among the most widely used of all construction materials. Such materials contribute more income to the world's economy than all other mineral resources combined, with the exception of mineral fuels such as petroleum and coal.

Nonmetallic Minerals and Rocks

Nonmetallic minerals and rocks come in a broad variety of types and sizes. They are used for construction purposes ranging from foundations of buildings to insulation to road base.

Building and Dimension Stones. Many different sizes of stones are widely used as facing (ornamental or protective) stones for fireplaces, houses, and other buildings. Specially shaped blocks and slabs of rock called dimension stones are used for foundations, roofs, **veneers** on walls, flagstones, tiles, curbs, and monuments. Almost all such stones are cut and shaped, and their surfaces are dressed, or finished, by grinding or polishing. Granite and granitelike rocks, limestone, sandstone, marble, and slate make up over 90 percent of the dimension stones used in the United States. Although concrete has replaced dimension stone in many buildings in recent years, some experts predict that builders will return to using dimension stone because it requires less energy to produce than cement, steel, or aluminum.

veneer thin sheet of material used as a protective or ornamental facing on a wall or piece of furniture

Crushed Stone. Rock that has been crushed, ground, or broken into small chunks is called crushed stone. A large percentage of crushed stone is used as

aggregate—filler that is mixed with asphalt or cement to give it greater strength. Crushed stone is also commonly used as a base or fill for roads and railroad beds, as riprap (chunks of rock for **jetties** and other shoreline protection structures), and as granules, or particles, that are combined to make roofing or flooring. Most crushed stone consists of limestone and dolostone, but it can also include materials such as basalt, sandstone, marble, and slate. Crushed stone is mainly quarried from open pits, although it is mined in some areas. Concrete is also sometimes crushed and recycled as "crushed rock." Crushed stone is generally used near where it is produced because its low unit value does not make it profitable to transport or export.

Other Nonmetallic Minerals and Rocks.   Sand and gravel are widely used as base or fill for roads and aggregate for asphalt and cement. Most of the sand and gravel produced in the United States is used for these purposes. These materials are also used as roofing granules and in decorative walkways and walls. Sand consists largely of quartz grains; gravel consists of sand grains plus pebbles of rocks such as quartzite that resist **weathering** and **erosion** relatively well. Like crushed stone, sand and gravel have a low value per unit and are typically used near their sources.

The term *asbestos* is used to refer to a group of silicate minerals that separate easily into strong, thin fibers. The fibers are flexible, heat resistant, and chemically **inert,** and they are used by manufacturers of "fireproof" products such as roofing and siding shingles, floor tiles, caulking, and insulation. Since the 1970s, the United States and other countries have banned, reduced, or phased out some uses of asbestos because of possible health risks. Long-term exposure to asbestos or asbestos dust may lead to certain lung diseases, including lung cancer. However, some uses of asbestos, especially those in which the asbestos is sealed inside a container, are likely to continue.

Asphalt, also called tar, is a material widely used to produce asphalt concrete for paving roads. Asphalt may occur naturally in lakes or seeps (places where water or petroleum comes up from below ground), but most of it is made from distilled petroleum. Asphalt concrete is made by combining asphalt with an aggregate of sand, gravel, or crushed rock. Asphalt is also used for tar paper, shingles, and pipe coatings.

Materials Manufactured from Rocks

In addition to the rocks themselves, many other construction materials are manufactured from nonmetallic minerals and rocks. These include cement, ceramics, plaster, and glass.

Cement.   The most commonly produced cement, Portland cement, is made from limestone and shale or related rocks. The rocks are crushed and combined according to a specific formula to produce the powder that is sold as cement. When cement is mixed with water and an aggregate in the right proportions and allowed to cure, it forms concrete. For many construction purposes, concrete is reinforced with metal bars. Portland cement can also be modified to produce cements with special properties, such as high resistance to acids.

Ceramics.   Ceramics, such as brick and tile, have clay or shale as their main ingredients. Moist clay is molded into the desired shape and then either dried in the sunlight or heated in a kiln to make it hard and rocklike. For production,

jetty (*pl.* jetties)   structure built out into the sea, a lake, or a river to counteract the effects of tides or currents

weathering   physical and chemical breakdown of rock that is stationary on Earth's surface

erosion   wearing away of land by wind and water

inert   exhibiting little or no chemical activity

the clay or shale may be crushed and blended before it is shaped and dried. The reddish orange color of many bricks comes from the iron present in the clay, but coloring agents can be added to iron-free clays to produce other colors. Brick is mostly used to face buildings, but some is used for walkways and patios. Tile is used for floors, walls, roofs, and drainage pipes.

Plaster and Glass.   Plaster is any material used to coat and modify the outward appearance of various surfaces, but the term almost always refers to plaster of paris. The main raw material in plaster of paris is gypsum, which is crushed, heated to remove water, and pulverized to make a powder. When this gypsum powder is mixed with water, it forms a paste that hardens into a fine-grained mass of gypsum. Almost all plaster is used to make prefabricated wall boards such as drywall, a layer of plaster sandwiched between two sheets of heavy paper.

Over the last 200 years, glass has been used in ever-increasing amounts as a construction material. It was probably first used by the ancient Romans, who used pieces left over from mosaics to decorate the walls of their houses. Particularly since World War II, glass has come to dominate the exteriors of many buildings, especially high-rise apartment and office buildings. It is also used in the form of fiberglass, which is the main component in many insulation materials and plumbing ducts. SEE ALSO CERAMIC MATERIALS; COAL; FOSSIL FUELS; GLASS; MINERALS; PETROLEUM; ROCKS.

Continental Crust

mantle   region of the Earth between the molten core and the outer crust

The Earth is composed of five basic layers: the inner core, the **mantle**, the crust, the hydrosphere, and the atmosphere. The crust consists of oceanic crust and continental crust. Oceanic crust is cooled magma (molten rock from the mantle). Continental crust is a mix of rocks—such as basalt, granite, granite gneiss, and peridotite—brought together by complex movements within the Earth's lower crust and mantle.

Crustal Movement.   Because the rock that makes up the crust is less dense than the mantle, it floats atop the mantle like ice floating on water. Both the continental and the oceanic regions of the crust rest on plates—large segments of crust and upper mantle that move slowly and often collide with one another. When an oceanic plate collides with a continental plate, the denser mantle

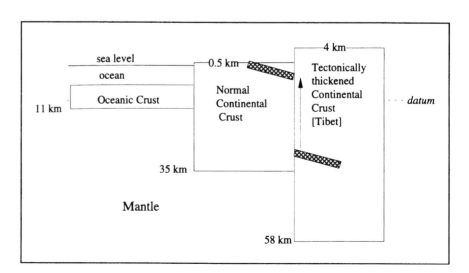

Evolution of crustal thickness—deep-seated rock rises toward the surface as mountainous areas are eroded down.

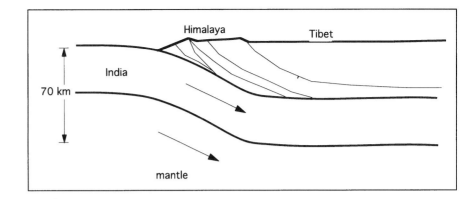

Collision of one continental plate with another, where one of the continents is forced beneath the other.

portion of the oceanic plate is pushed under the continental plate, while less dense material from the oceanic crust is compressed against the continental crust. Over millions of years, the crustal material piles up at the shoreline where the two plates meet and forms coastal mountain ranges.

When continental plates collide with each other, two outcomes are possible. In the first case, their edges may collide, forcing crustal material to pile up and form a mountain range. The depth of the crust increases where the collision occurs, and the plates stop moving. The Pyrenees Mountains in Europe are an example of this type of collision. In the second case, one of the continental plates is forced under the crustal portion of the other. The material on the top plate is raised up and the crust thickens, but the plates continue to move. This movement can eventually force the crust to be raised up to a height of 4 to 5 km. The mountain range known as the Himalaya and the high plateau of Tibet are results of this type of actively continuing collision between India and Asia, where the forces of **erosion** cannot break down the crust as fast as the colliding plates build it up.

erosion wearing away of land by wind and water

The Stable and Unstable Crust.   On most continental plates, the crust averages about 35 km thick. However, where plates collide to form mountain ranges, the crust at the site of the collision can become much thicker. In Tibet, for example, the crust is nearly 70 km thick—a highly unstable situation. Temperatures at the depths reached by lower parts of such thickened crust are high enough to cause crustal rocks to melt and flow. The lowered density of the melted rock allows it to rise to the surface and the remaining, more solid rock is then able to move laterally, or sideways. At the same time, the rocks supporting such mountains are subject to erosion and collapse (for instance, due to landslides), and eventually the surface is eroded down to near sea level. As a result, the crust returns to its more stable, 35-km thickness, with formerly deep-seated rock now at or near the surface. SEE ALSO EARTH'S CRUST, HISTORY OF; LANDSLIDES AND ROCKFALLS; MAGMA; MANTLE; MOUNTAINS AND MOUNTAIN BUILDING; PLATE TECTONICS; TECTONISM, ACTIVE; WEATHERING AND EROSION.

Continental Drift

See Plate Techtonics.

Continental Margins

Continental margins are those regions where continental crust meets oceanic crust. They are an object of considerable interest to geologists because they contain a great deal of information about Earth's history, including changes in climate, the composition of the atmosphere, sea level and water chemistry,

Continental Margins

marine life, and the movement of **tectonic plates.** Continental margins also contain much of Earth's petroleum resources and are critical habitats that make up its most important fishing grounds.

Composition of Continental Margins

Although they are covered by ocean water, the margins are considered geologically part of the continents. They are composed of three basic parts: the continental shelf, the continental slope, and the continental rise. The continental shelf is a shallow, gently sloping platform that extends into the ocean from the shoreline. At its edge is the continental slope, a surface that dips steeply to the ocean floor. The continental rise is a layer of **sediment** that has been worn away from the continent and has settled on the ocean floor, where it partially overlaps the seaward edge of the continental slope.

Types of Margins

Two types of continental margins exist: passive margins and active margins. These two types of margins are formed in different places, by different processes that produce different geological features.

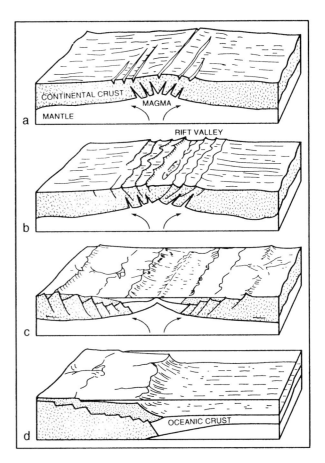

Stages in the breakup of a continent, the development of Atlantic-type passive margins and ocean floor.

a. Thermal uplift of a broad region as magma intrudes.

b. Uplift of mantle with faulting and formation of a rift valley.

c. Formation of new oceanic crust with faulting, uplift, and erosion of rift shoulders and cooling and subsidence of continental margins.

d. Crustal thinning through faulting and cooling, with subsidence and sedimentation along the newly formed margins.

continental plate   large segment of Earth's crust and uppermost mantle (region just below the crust) that supports a major landmass

magma   molten rock from deep within the Earth

mantle   region of the Earth between the molten core and the outer crust

fault   break or crack in the Earth's rock layers that creates a surface along which the rocks slide

oceanic plate   large segment of Earth's crust and uppermost mantle (region just below the crust) that supports an area of ocean bottom

Passive Margins.   Passive margins form in the interior of an existing **continental plate** when **magma** surges up from the upper **mantle** into the continental crust. As the heat from the magma expands the crust, the plate is rifted (broken apart) and rises up to form an elevated plateau. The Colorado Plateau in the western United States was created as a result of this type of rifting action. The rifting produces additional **faults** that allow more magma to intrude (push up) into the crust, which becomes thinner and is stretched apart by the action of the magma. The results of such action can be seen in East Africa, where columns of heated rock rising from the mantle (called mantle plumes) have broken and stretched the crust, creating rift valleys with high, steep walls.

If a rift forms where there is sufficient water, it fills to form a new lake or shallow sea and creates the passive margin. The Red Sea is an example of newly formed ocean crust and youthful passive margins. As the seafloor continues to spread, the continental crust on either side moves farther from the source of heat, cools down, and subsides (sinks down), forming a basin along the margin. Sediment eroded from the adjacent continents collects in such basins. In some places off the Atlantic coast of North America, the sediments are 5 to 6 km thick. As the basin fills with sediment, the crust beneath it subsides further, so the basin may never be completely filled in.

The Atlantic Ocean margins of North America, Africa, and Europe are passive margins that began to form about 200 million years (Ma) ago when the interior of a large continent, called Pangaea, began to break up. It split into two plates called Laurasia (North America, England, western Europe, and Ukraine) and Gondwana, or Gondwanaland (Australia, India, Africa, South America, and Antarctica). The rift that caused them to split, known as the Mid-Atlantic Ridge, now lies on the ocean floor. Although active faulting continues along the rift, seafloor spreading has moved the two plates thousands of kilometers away from the Mid-Atlantic Ridge, so the continental margins are considered tectonically quiet.

A passive margin has a broad continental shelf, a gentle continental slope, and a more pronounced continental rise. Although the seaward edge may be marked by deep valleys where currents have caused erosion, sediments continue to deposit in the subsiding basins of passive margins. This process produces a coastal region consisting of wide beaches, offshore barrier islands, and well-washed sand.

Active Margins.   Active margins form primarily along the boundaries of plates that are actively converging (moving toward each other) and where subduction is occurring (one plate is being pushed underneath another). Active margins mark the continents that border the Pacific Ocean. Subduction of **oceanic plates** along the rim of the Pacific has produced a wide variety of continental margins, such as the Andes Mountains in South America and the seaside cliffs on the northwestern coast of North America.

On South America's Andean coast, subduction of the Nazca oceanic plate caused it to melt partially. This melting led to the formation of magma that intruded into the continental crust just above, producing the extensive volcanoes and steep peaks of the Andes. This process began about 140 Ma ago when South America broke away from Gondwana and began moving westward. The active margin and the Andes formed about 100 Ma ago. This type of active margin is known as a convergent margin.

Some plates contain microcontinents and island arcs. Microcontinents (such as New Zealand) are formed when a segment of a continent separates from the mainland; island arcs (such as Japan) are formed by a series of volcanoes that are created when one oceanic plate is subducted beneath another. Microcontinent and island arc material is less dense than oceanic crust; it is too buoyant to be subducted. When such a plate collides with a continental plate, the microcontinents and islands accrete (pile up) along the edge where the two plates meet. The resulting continental margin is called an accreted margin.

Geologists believe that about 180 to 80 Ma ago there were many island arcs and microcontinents in the eastern Pacific. As the North American plate moved westward during the breakup of Gondwana, the Pacific Ocean plate subducted beneath the continental plate to its east, and these islands and microcontinents accreted along North America. If California's San Andreas Fault continues to be active, geologists predict that the Los Angeles area will separate from the mainland and move northward. In about 10 Ma, it will pass San Francisco and in about 60 to 80 Ma, it will accrete along the Alaskan coast. Accreted margins such as this form a mosaic of terranes (rock formations) of various rock types and ages, separated by faults.

Active margins have much narrower continental shelves than passive margins, and their continental slopes are usually much steeper. The underwater **topography** of active margins is also more diverse, including island arcs and deep-sea trenches. The coast along an active margin is typically uplifted by the subduction of the adjacent plate, forming terraces and cliffs that are cut by the ocean's waves. If the coast were not uplifted in this way, wave erosion would eventually cut the coast down to a flat plain. SEE ALSO ANCIENT SUPERCONTINENTS; CONTINENTS, EVOLUTION OF; EARTH, STRUCTURE AND COMPOSITION OF; MAGMA; MANTLE; MOUNTAINS AND MOUNTAIN BUILDING; PLATE TECTONICS; RIFTING OF THE CRUST; SEDIMENTS AND SEDIMENTARY ROCKS; TECTONISM, ACTIVE; WEATHERING AND EROSION.

topography physical features of a place or region

Continents, Evolution of

tectonic plate large segment of Earth's crust and uppermost mantle (region just below the crust) that moves as a unit over the Earth's surface, floating on a partially molten layer of rock below

The entire surface of the Earth is covered with a relatively thin layer of rock known as the crust. However, the crust that makes up the continents is very different from the crust beneath the oceans. Continental crust is much thicker (averaging 38 km thick) than ocean crust (averaging about 7 km), is less dense, and is much older. The oldest ocean crust is about 160 million years old, whereas the age of continental crust averages 1.7 billion years and is as old as 4 billion years in some places. Various geological processes involving **tectonic plates** operate continuously, creating and destroying continental and oceanic crust. These processes have caused the continents to evolve over time, and they continue to alter continental size, shape, and position on the earth.

Explaining Differences in Types of Crust

magma molten rock from deep within the Earth

mantle region of the Earth between the molten core and the outer crust

oceanic plate large segment of Earth's crust and uppermost mantle (region just below the crust) that supports an area of ocean bottom

The differences between oceanic and continental crusts are mainly due to the fact that ocean crust is "recycled" (created and destroyed) much more rapidly than continental crust. The oceans contain volcanic ridges where **magma** from the **mantle** surges up into the crust, pushing apart the seafloor and creating new crust (a process called seafloor spreading). As new ocean crust moves away from these spreading ridges, the layer of mantle rock directly below the crust cools and stiffens, becoming part of the **oceanic plate.** About 40 million years after it is formed, the plate becomes denser than the hotter mantle on which it rests. Consequently, it begins to sink into the mantle along deep-sea trenches. This process is called subduction.

continental plate   large segment of Earth's crust and uppermost mantle (region just below the crust) that supports a major land mass

sediment   soils, rock particles, and other materials that are deposited over time and make up the ground, whether on dry land or at the bottom of a body of water

Continental plates, however, resist subduction because their thick crust has a relatively low density. As a result, they "float" on top of the mantle and cannot be subducted while intact, although smaller pieces of the continental crust can be recycled. For example, an ocean plate that is being subducted beneath a continental plate may scrape off continental rock, or continental **sediments** may be washed into the ocean and later carried into the mantle by subducting ocean plates. Because subduction has a greater impact on ocean plates than on continental plates, virtually all of the ocean crust that currently exists will have been recycled in 200 million years, while over 97 percent of existing continental crust will still be intact.

Processes That Create Continental Crust

There are four main processes by which continental crust is created: trench accretion, arc magmatism, plume magmatism, and rift magmatism. These processes work together in various combinations.

Trench Accretion.   A deep-sea trench forms where an oceanic plate is subducted beneath a continental plate. Trench accretion occurs when material is scraped off the subducting plate and accretes (builds up) onto the overriding plate. The accreted material is often continental sediment that has built up in the trench, in which case no net continental growth occurs. However, at some trenches, volcanic material from the mantle that has built up on the oceanic plate accretes onto the continental plate and becomes new continental crust.

Arc Magmatism.   Where an oceanic plate is subducted into a deep-sea trench, water is released into the hot mantle material beneath the overriding plate. This water lowers the temperature at which the mantle material melts, producing magma. The less dense magma then rises into the crust of the overriding plate, leading to volcanic activity at the surface. As the oceanic plate continues to subduct, more water enters the mantle, and the formation of magma continues. Because most sea trenches where subduction occurs are arc-shaped, the magma from the resulting surface volcanoes is usually deposited in the shape of an arc. For this reason, the process is called arc magmatism. When such an arc occurs at the meeting of two oceanic plates, it forms what is known as an island arc, such as Japan. An island arc may eventually collide with the margin of a continent and be added to the continent by accretion. The island of Taiwan, for example, is currently in the process of accreting to the southeastern coast of China.

Plume Magmatism.   At the base of the mantle, some 2,900 km beneath the Earth's surface, narrow, hot jets of material (called plumes) are created by heat from the Earth's outer liquid iron core. As plume material nears the surface, it begins to melt, creating magma that collects within or upon the crust and increases its thickness. This process is called plume magmatism, and it is currently occurring in the Hawaiian Islands and Yellowstone National Park.

Plume magmatism is most effective at thickening the crust on plates that are relatively stationary. The African plate, for example, is currently resting above a set of plumes. As a result, Africa has many volcanoes and an unusually thick crust. When a plume occurs close to a seafloor spreading ridge, as in Iceland, it produces an oceanic crust of about three times the normal thickness—20 km or more. Plateaus on the seafloor may occur as a result of plume magmatism, and these may eventually be added to continents by trench accretion.

Rift Magmatism. Rift magmatism occurs when horizontal stress stretches apart an area of the crust (a process called necking), reducing pressure in the mantle just below. The stress may be caused by any of a number of factors, such as the sinking of a plate into a trench or the collapse of a continental plateau into the mantle. As the thinning crust releases pressure, mantle material melts more easily and rises into the rift, or crack, that has been created in the crust. The amount of melting depends upon the temperature of the mantle and the rate at which the necking occurs. When extreme necking caused by the breakup of a continent occurs near a mantle plume, about 25 km of crust may be added to the continental margin.

Rifting often occurs in areas affected by arc magmatism and results in the formation of seas near continental margins. The Sea of Japan is an example of a marginal sea created by rift magmatism. When a magmatic arc rifts, water-release and pressure-release factors combine to increase the rate of mantle melting and crust formation. However, this process tends to produce thicker oceanic crust, rather than thicker continental crust.

Composition of Continental Crust

The various types of magmatism that contribute to the formation of continental crust bring to the surface molten rock that is similar in composition to basalt lava. However, most of the magmatic rocks in the upper continental crust are granitic (granitelike). They contain higher amounts of silica, alumina, and alkalis than basalt lava, and lower amounts of magnesium, calcium, and iron.

Most granites are produced by secondary melting, a process that occurs when rocks from the deep crust are heated by molten rock rising from the mantle. These granitic melts then rise through the crust. As a result, the upper crust becomes more granitic, and granitic material is removed from the lower crust. Logically, then, the lower crust should be basically basaltic. In order to find out if this is true, scientists must study relatively complete sections of lower crust. The ideal locations are places like the Alps and the Himalaya, where plates have collided and the lower crust has later been exposed by the combined effects of crustal thickening and erosion. Scientists can also learn about the lower crust by studying xenoliths, lower crust fragments brought to the surface by volcanic explosions.

From the study of exposed lower crust and xenoliths and other observations, geologists have determined that the composition of the lower crust is not primarily basaltic. There are two possible explanations for this. First, some basaltic material in the lower crust may have sunk back into the mantle because of its greater density. Second, the processes that form crust may have been different when the Earth was younger, and these may still be represented because continental crust is recycled so slowly. To evaluate the second possibility, geologists have studied the oldest regions of the crust, known as the Archean shields. These sections of crust, formed over 2 billion years ago, are relatively well preserved because the mantle upon which they sit is quite stable. This stability is due to the fact that most of the material likely to become molten from those portions of the mantle was removed when the crust first formed. The areas that have been most intensively studied are in central Canada, southern Africa, and western Australia.

Wandering Continents

Continents evolve so slowly that they appear never to change. However, since they first formed billions of years ago, the continents have broken up and recombined repeatedly. Scientists believe that at one time, all of the continents were joined together in a single landmass called Pangaea. About 150 million years ago, Pangaea began to break up to form separate continents. Some of these drifted apart and became the continents of the modern globe. Others collided with larger continents and became part of them, leaving mountain ranges as evidence of their impact. The mountains of the Himalaya are the result of such a collision between India and Asia.

These portions of the crust give us insight into the processes that created them. Deformed versions of oceanic plateaus, volcanic island arcs, rift volcanoes, and trench sediments represent places where accretion has clearly occurred. These areas contain sheets and plugs of granitic material that may be the result of arc and rift magmatism at continental margins. Chemical evidence found in the granitic material, however, suggests that it was created partly by the melting of oceanic crust. Scientists believe that the source of these early arc magmas was silica-rich molten rock from the oceanic crust. Later in Earth's history, basaltic molten rock from the mantle may have become the main source of new continental crust. This may explain why the lower crust, which is older, contains less basaltic material than might be expected.

The explanation for the composition of lower continental crust remains unclear. Selective recycling processes (sinking basalt), changes in crust formation, or both may be involved. To resolve this issue, scientists need to conduct more studies of modern and ancient continental crust. They also need a more precise understanding of the processes that create and destroy continental crust. SEE ALSO CONTINENTAL CRUST; CONTINENTAL MARGINS; EARTH'S CRUST, HISTORY OF; MAGMA; MANTLE; PLATE TECTONICS; RIFTING OF THE CRUST; SEDIMENTS AND SEDIMENTARY ROCKS; TECTONISM, ACTIVE; VOLCANOES; WEATHERING AND EROSION.

Corals

See Fossils, Colonial Invertebrate.

Core, Structure of

See Earth, Structure and Composition of.

Cosmic Dust

See Interplanetary Medium.

Cosmology

Cosmology is the branch of **astrophysics** that deals with the origin, structure, and evolution of the universe. The current picture of the universe began to develop in the early 1900s, has been enriched and modified by the work of many scientists, and continues to unfold.

astrophysics science that deals with the physical characteristics of stars and stellar events or conditions that can be observed

gravitational field area influenced by gravity

astronomer scientist who studies space and the various bodies it contains

Einstein and Relativity. In 1905 Albert Einstein published three of the most important physics papers of the 1900s, including his "Special Theory of Relativity." He followed these in 1916 with another paper, "The Foundation of the General Theory of Relativity," which dealt with the motion of bodies in a **gravitational field.** To explain his theory, he created a series of "field equations"—mathematical equations that could describe the shape of the universe based on its distribution of matter and energy. Einstein believed that these equations could allow scientists to explore the origins of the universe and predict its future.

However, Einstein did not know the actual distribution of mass and energy in the universe. (**Astronomers** were still developing tools to measure the distances to other stars and galaxies and to determine whether, and how fast, those objects moved.) To solve his equations, he used a reasonable assumption called the "cosmological principle," which states that the universe "must appear the same for all observers." This means that the universe is homogeneous (the same everywhere) and static (unchanging over long periods of time).

From 1917 to 1927, several scientists presented alternative solutions to Einstein's field equations. They suggested that the universe might be not static but dynamic—changing in size or shape over time. These solutions received little attention until 1931, when astronomer Arthur Eddington asked Britain's Royal Astronomical Society to reprint a solution worked out by Georges Lemaitre in 1927. By the early 1930s, astronomical evidence strongly suggested that nonstatic solutions to Einstein's equations might be closer to reality than static solutions.

The Expanding Universe. Evidence that the universe was expanding, not static, was actually presented long before Lemaitre's paper was reprinted. In 1914 an American astronomer, Vesto Slipher, determined that 14 spiral **nebulae** were moving away from our galaxy at high speeds. He calculated these speeds by measuring the "redshift"—the difference between color patterns of light observed in distant galaxies and patterns given off by chemicals on Earth. (When an object is moving away from an observer, the light it gives off appears to shift towards the red end of the **spectrum,** or range of visible light.) However, the full significance of Slipher's discovery would not become clear for another 15 years.

In 1924 Edwin Hubble developed a technique for measuring the distances to certain stars based upon their brightness as perceived from Earth. By 1929 the redshift had been determined for 46 galaxies, and Hubble had used his technique to measure the distances to 18 of them. He found that the farther away a galaxy was, the faster it was moving away from the observer. Einstein was impressed by Hubble's findings, and in 1930 he traveled to Mount Wilson Observatory in California to meet with Hubble. Based on their discussion, Einstein changed his mind and came to agree with the idea of an expanding universe.

Big Bang Theory. Applying Einstein's equations to an expanding universe, one can work backward in time to a point when the universe was very dense and packed together in a very small space. From this point, the universe expanded explosively, so this type of theory about the origin of the universe is called a "big bang" theory. However, one problem that scientists have with any big bang theory is that it implies a specific beginning for the universe. Since the scientific method is based on the concept of cause and effect, this implication presents a dilemma. What caused the universe to begin? Many scientists feel that this idea of a "first cause" is easier to approach from a religious perspective than through scientific investigation.

Another question posed by the idea of an expanding universe is whether the expansion will ever stop. The answer depends upon the average **mass** density of the universe—that is, the total mass of everything in the universe divided by the total volume of the universe. If the total mass density is greater than a certain "critical density," the universe will stop expanding and eventually collapse upon itself. This possibility is called a "closed" universe. If the total is less than the critical density, then the universe is "open" and will continue to expand forever. If the universe is (and remains) exactly at its critical density, then it is a "flat" universe: its rate of expansion will slow until it approaches zero, but it will never reach a point where it reverses so that the universe begins to collapse.

The critical density of the universe has been calculated according to the standard big bang theory. However, as of the late 1900s, observations suggested that the average mass density of the visible matter in the universe is only about 3 percent of this critical density. This calculation means either that

nebula (*pl.* nebulae) body of gas or dust within our galaxy; also, distant galaxy that appears as a cloud of light

spectrum range of visible light

The Raisin Bread Universe

Edwin Hubble showed that the farther away a galaxy is, the faster it is moving away from the observer. To understand why, imagine the universe as a lump of bread dough containing raisins, which represent the galaxies. At first, the raisins are close together. However, as the yeast in the dough causes it to expand, the raisins grow farther apart. No matter which raisin you choose as the starting point, all the other raisins will move away from it as the dough expands. Because all of the dough expands at the same rate, the more dough there is between two raisins, the faster they move apart.

mass amount of material that causes an object to have weight when it is in an area influenced by gravity; commonly measured in kilograms

the origin of the universe was a one-time event—and that the universe will continue to expand—or that the universe contains unobserved mass that cannot presently be detected or measured. Scientists have speculated that this so-called "missing mass" may be present in black holes, subatomic particles, or some other unidentified form. The best evidence for the existence of some form of "dark matter" comes from observation of the way many spiral galaxies rotate, which strongly suggests that they contain hidden matter.

A U.S. Naval Observatory photograph of the Veil Nebula.

thermodynamics branch of physics that deals with various forms of energy

quantum mechanics branch of physics that deals with the structure and behavior of atoms and subatomic particles

Crustaceans

Currents

Cuvier, Georges

1769–1832
French paleontologist

natural history systematic study of natural objects, especially in their natural settings (includes physical and life sciences)

The Steady State Theory.   The big bang theory was not immediately accepted by all scientists. In 1948 Hermann Bondi and Thomas Gold proposed the steady state theory, which argues that as the universe expands, new matter is automatically created to fill it. However, this theory violates the laws of **thermodynamics,** which state that the sum total of energy and mass in the universe remains constant over time.

That same year, George Gamow reasoned that if the big bang had occurred, it should have left traces of a particular type of radiation. If such radiation could be measured, it would support the big bang theory and disprove the steady state theory. Such radiation was detected in 1965, when Arno Penzias and Robert Wilson were testing a new antenna that had been created to observe satellites; they noticed background radiation that seemed to come from the whole sky. The temperature of the radiation matched predictions made the year before that estimated the current temperature of radiation left over from the big bang. Penzias and Wilson later received the Nobel Prize for their work.

The Inflationary Universe and Anisotropies.   A new twist on the big bang theory, called the "inflationary universe" theory, gained attention during the 1990s. This theory suggests that just after its origin, the universe existed in a state called a false vacuum. In this state, rapid expansion would naturally occur, and all matter would be created out of the energy produced by the vacuum. This theory is quite complex and requires an understanding of **quantum mechanics** to appreciate fully.

One problem not addressed by either the standard big bang or the inflationary universe theory is the origin of anisotropies, very slight variations in the temperature of the background radiation. These variations provide a theoretical explanation of how clumps of matter came together to form galaxies and clusters of galaxies, and their existence is supported by the 1989 data from the Cosmic Background Explorer (COBE) satellite. Just how anisotropies came about in the early universe, however, is a question yet to be answered by cosmologists. SEE ALSO ASTRONOMY, OBSERVATIONAL; GALAXIES; HUBBLE, EDWIN POWELL.

See Arthropods.

See Coastal Erosion and Deposition; Oceanography.

Born in Montbéliard in eastern France, Georges Cuvier originally trained in Stuttgart, Germany for a career in either law or the ministry. He became interested in **natural history** while working as a tutor for the family of Comte d'Héricy, who lived in Normandy, France. While in France, Cuvier was introduced to prominent **naturalists** in Paris, and in 1795 he became an assistant at the Museum of Natural History in Paris as well as a member of the French Academy of Sciences. He rose quickly through both academic and governmental ranks, holding many titles and positions. Cuvier was made a baron by King Louis Philippe in 1831.

By the early 1800s, Cuvier had become a major figure in the scientific community throughout the western world. His specialty was comparative anatomy, especially of living and extinct **vertebrates,** and he is sometimes

naturalist student of natural history (the systematic study of natural objects, especially in their natural settings)

vertebrate any animal that has a backbone or spinal column

comparative anatomy science that deals with comparisons of the structures of different plants and animals

paleontology science that deals with prehistoric plants and animals through the study of fossils and other remains

species narrowest classification, or grouping, of organisms according to their characteristics; members of a species can reproduce only with others of that group

extinct having died out completely, leaving no direct descendants (refers to a specific grouping of organisms)

evolution changes in groups of related organisms occurring over time

Georges Cuvier.

considered the founder of **comparative anatomy** as well as the father of vertebrate **paleontology.** Cuvier paid particular attention to the relationships among various body parts in the same animal. Given a jumble of bones of more than one fossilized **species,** Cuvier could sort the bones into the proper species. It was sometimes suggested that from a single bone, he could reconstruct an entire animal.

Cuvier was also interested in the classification of animals, both living and **extinct.** Although especially interested in fossil mammals, he also studied reptiles and gave the name pterodactyl ("wing finger") to a type of extinct "flying reptile." With the French geologist Alexandre Brongniart, Cuvier studied, analyzed, and documented the rock layers and fossils of the Paris region.

Although Cuvier proposed his own system for classifying organisms and recognized the close similarities among many species, he rejected the idea of **evolution.** Cuvier believed that every species was adapted to one particular way of life, and any significant change in a species would upset the relationships between the functioning parts of the organism as well as between the organism and its environment. Therefore, Cuvier argued that organism could not change significantly or evolve over time.

Cuvier's studies of fossils established that there were fossil forms that did not resemble any living organisms. He concluded that many fossil animals had become extinct. His contemporary, the French naturalist Jean-Baptiste Lamarck, rejected this idea. Lamarck argued that "extinct" animal species had actually evolved gradually into other living species.

Cuvier's work laid the foundations for the theory known as catastrophism—the idea that massive and sudden geological changes and upheavals periodically occur in Earth's history. Although Cuvier did not believe that the new organisms that appeared after a catastrophe or upheaval were created by God, his work contributed to this idea of miraculous creation, sometimes called "scientific creationism." Both theories opposed the ideas of Lamarck and Charles Darwin, who believed in the relatively slow and gradual evolution of species. SEE ALSO BRONGNIART, ALEXANDRE; DARWIN, CHARLES ROBERT; FOSSILIZATION AND THE FOSSIL RECORD; LAMARCK, JEAN-BAPTISTE; LIFE, EVOLUTION OF; PALEONTOLOGY; PALEONTOLOGY, HISTORY OF; SCIENTIFIC CREATIONISM.

Daly, Reginald Aldworth

1871–1957
Canadian geologist

mantle region of the Earth between the molten core and the outer crust

petrology branch of geology that deals with the origin, structure, composition, changes, and classification of rocks

magma molten rock from deep within the Earth

The Canadian geologist Reginald Aldworth Daly provided new insights into the composition and structure of Earth's crust and upper **mantle.** He made important contributions to the **petrology** of igneous rocks (made from **magma**) and other aspects of geology. Daly was born in the small town of Napanee in the Canadian province of Ontario. While studying for a bachelor of arts degree at Victoria College (now the University of Toronto), Daly showed no special preference in his studies. It was not until his extra year at Victoria, while pursuing a bachelor of science degree, that Daly became very much interested in geology. He entered graduate study at Harvard University and taught there from 1898 to 1901.

Between 1901 and 1907, Daly conducted an ambitious field study for the Canadian Geological Survey of the geology along the western border between Canada and the United States. Helped by only one assistant, he mapped the entire area's geology and collected thousands of rock specimens and photographs. His numerous observations and detailed analyses provided an important geological overview of that enormous area.

Returning to the United States, Daly taught at the Massachusetts Institute of Technology from 1907 to 1912 and at Harvard University from 1912 to 1942.

During this time, he wrote a number of papers and books based on his field-work in Canada and the United States.

In 1914 Daly wrote his first textbook, *Igneous Rocks and Their Origin,* in which he described the chemical and physical properties of igneous rocks. From his fieldwork, he developed important concepts about the formation, chemistry, and physical properties of these types of rock. Daly considered theory to be the basis of practical geology, and he believed that the data to support geological theories must come from large-scale geologic mapping. A superb lecturer and popular teacher, Daly presented his ideas clearly and concisely to his students, who were known to applaud at the end of his lectures.

In 1932 Daly initiated a program of research in **geophysics** and experimental geology. Over the course of 40 years, this project produced many significant contributions to our knowledge of Earth's interior. He also helped establish a **seismological** station in Massachusetts, near Harvard University.

Daly's principal influence was through his university lectures, his participation in scientific societies and activities, and his many writings. In addition to scholarly works, he wrote a number of books aimed at a general audience. Filled with illustrations and data, these books did much to popularize the field of geology. SEE ALSO EARTH, STRUCTURE AND COMPOSITION OF; IGNEOUS ROCKS.

geophysics science that deals with the physical characteristics and structure of the Earth

seismological related to seismology, the science that deals with earthquakes

Dana, James Dwight

1813–1895
American geologist

mineralogy science that deals with minerals and their properties

paleontology science that deals with prehistoric plants and animals through the study of fossils and other remains

James Dwight Dana.

One of the most prominent American scientists of the 1800s, James Dwight Dana made important contributions to the fields of geology, **mineralogy,** and **paleontology,** as well as to the general development of American science.

Born in Utica, New York, Dana was raised in a strict Christian family. At an early age, he collected and was fascinated with minerals, plants, and animals. In 1830 Dana entered Yale University (then called Yale College) to study with the leading science teacher of the day, Benjamin Silliman. Dana followed a traditional curriculum of theology, philosophy, science, and other subjects. A serious student, he spent much of his leisure time studying Silliman's mineral collection and adding to his own collections.

During the early 1800s, the field of American geology was just developing. The greatest controversy of the period concerned the origin of rocks; conflicting views were held by the scientists Abraham Gottlob Werner and James Hutton. Religion was a major issue in discussions of geology at the time, and arguments raged over the creation of the Earth by natural or divine forces. Dana, like many of his contemporaries, tried to interpret geologic concepts to fit his own religious beliefs.

Dana traveled throughout the Mediterranean region in 1833 and 1834, working as a mathematics teacher to sailors aboard the USS *Delaware.* Experiences from this and other travels laid the groundwork for some of his major studies. They also provided the basis for his views on geology, especially concerning volcanoes and coral reefs. Throughout his career, Dana corresponded with Charles Darwin, who had developed the concept of the upward growth of corals with the rise of sea level. Dana's work on the evolution of coral growth and volcanic formations largely, though not entirely, built on and supported Darwin's idea.

From 1838 to 1842, Dana participated in the Wilkes expedition. A major event in American science, the expedition circled the world, conducting natural history surveys. The expedition served as a springboard for the development of Dana's career. He not only continued his study of minerals but also gathered data and observations that resulted in numerous scientific papers.

His experiences and writings made Dana a respected scientist, and his duties as editor of the *American Journal of Science* from 1846 into the 1880s helped expand his growing influence.

Dana's studies led to the development of the first great American theory about the geologic history of the Earth. Dana hypothesized that folded belts of mountains, such as the Appalachians, were formed as a result of subsidence (sinking) of different parts of the Earth as it cooled—an idea that later contributed to the theory of **plate tectonics**. Dana also published his views in his influential *Manual of Geology*, which played a critical role in the development of geology as a theoretical science as well as a descriptive one. Dana's most lasting contribution was a massive work, *A System of Mineralogy*, which became a standard reference work on the compositions and characteristics of minerals. Dana played a prominent role in the development of the U.S. Geological Survey and the National Academy of Science. SEE ALSO FAMOUS CONTROVERSIES IN GEOLOGY; GEOLOGY, HISTORY OF; HUTTON, JAMES; WERNER, ABRAHAM GOTTLOB.

plate tectonics theory used to explain continental drift (the movement of continents over the surface of the Earth) and other geological processes

Darwin, Charles Robert

1809–1882
British biologist
and geologist

natural history systematic study of natural objects, especially in their natural settings (includes physical and life sciences)

Darwin and the Galápagos Islands

The *Beagle* spent only a month in the Galápagos Islands, off the west coast of South America. It was Darwin's observations there, however, that helped him develop his theory on the origin of species. Darwin collected and examined 13 bird specimens, all of which proved to be finches, although they looked quite different from each other. Darwin concluded that these different species of finches had evolved from a single parent species. Their differences equipped them to live in various ways—thick bills were for cracking hard seeds and long thin ones for probing into blossoms, for example. Modern scientists regard isolated island areas such as the Galápagos as ideal laboratories for the study of how animals change, adapt, and evolve.

Charles Robert Darwin was one of the most influential and most controversial scientists who ever lived. He is best remembered as the biologist who provided overwhelming evidence that species of plants and animals change, or evolve, over millions of years. Although Darwin's evolutionary theory brought dramatic changes to the life sciences, his accomplishments as a geologist have received less attention.

Around the World. Darwin was born in Shrewsbury, England. In 1825 he began medical school in Edinburgh, Scotland. While studying medicine, he also pursued his interest in **natural history** and belonged to the university's biological and geological societies. Darwin eventually lost interest in medicine, and in 1828 he transferred to Cambridge University in England to study theology. Meanwhile, his interest in science grew, and he began discussing the field of geology with the Reverend Adam Sedgwick, a Cambridge professor and geologist. Darwin accompanied Sedgwick on a two-week field trip to Wales in 1831. Later that year, Darwin left England on a field trip that was to last much longer.

Darwin was the traveling companion of a naval captain during a five-year voyage around the world in the HMS (His Majesty's Ship) *Beagle*. Much of the journey was spent in the waters around South America—the *Beagle*'s official mission was to prepare sea charts of that region. The young naturalist explored wherever the ship landed.

Darwin noted fossil evidence of uplifted layers of the Earth's crust high in the Andes Mountains. He also experienced a dramatic earthquake in Chile. His notebooks of the trip contain four times as much geological information as biological data, although he eagerly collected thousands of samples of plants, animals, and insects.

Scientific Achievements. Upon his return to England, Darwin settled in London and began to publish his scientific observations. His account of his journey, in particular, is an outstanding work of both travel literature and science.

In 1839 Darwin married, and three years later, he moved to the country village of Down, where he spent the rest of his life. He built a scientific reputation with a steady stream of books based on his studies. He wrote, "It is very pleasant easy work putting together the framework of a geological theory, but it is just as tough a job collecting & comparing the hard unbending facts."

Charles Robert Darwin.

species narrowest classification, or grouping, of organisms according to their characteristics; members of a species can reproduce only with others of that group

In 1842 Darwin published his most important work in geology, *The Structure and Distribution of Coral Reefs.* Darwin's basic idea was that corals grow upward as a volcanic cone sinks and sea level rises, forming an atoll—a ring-shaped coral island surrounding a lagoon. Later both James Dana and William Davis accepted and expanded this pioneering work. Darwin continued to work on his geologic observations from the *Beagle* voyage until 1846, publishing a volume on volcanic islands and another on the geology of South America.

In 1859 Darwin published *On the Origin of Species,* the result of years of research and thought. He proposed that over long periods of time, minor variations in individual animals or plants sometimes allow those individuals to live longer, breed more successfully, and compete with other, less well equipped individuals. A process that Darwin called natural selection favored organisms with these beneficial variations, and through many generations, the variations became a new **species.** This radical theory openly defied the current scientific and popular belief that God created all life (most importantly, humans endowed with special gifts) and that the history of life as depicted in the Bible was complete and literally true.

During the latter part of Darwin's life, the controversy over evolution and humankind's link with the animal world overshadowed his valuable geological work on coral reefs. However, Darwin's biological and geological studies were parts of a unified picture of the world, not separate compartments of knowledge. As the scientific community and the public accepted the reality of evolution, they also accepted the concept of the great length of time required for evolution to operate. Together the concepts of evolution and geologic time changed people's view of the natural world and their place in it. SEE ALSO DANA, JAMES DWIGHT; DAVIS, WILLIAM MORRIS; FOSSILS, COLONIAL INVERTEBRATE; LIFE, EVOLUTION OF; SEDGWICK, ADAM.

Davis, William Morris

1850–1934
American geologist

erosion wearing away of land by wind and water

meteorology science that deals with the atmosphere, especially with regard to climate and weather

physical geography branch of geography that deals with the Earth's surface and the distribution, description, and nature of its land and water areas

geomorphology study of the Earth's surface and the forces that shape it

William Morris Davis is best known for his theories concerning long-term **erosion** and his system of describing landforms and how they change. Davis also contributed to the earth sciences by helping to popularize the use of block diagrams as a way of visualizing geologic ideas.

Born in Philadelphia, Davis earned degrees in geology and mining engineering from Harvard University. He assisted astronomer Benjamin A. Gould for several years in Argentina before returning to Philadelphia and then to Harvard, where he taught, researched, and published works in **meteorology,** geology, and **physical geography.**

Beginning around 1888, Davis worked out a system for analyzing landforms based on ideas formulated by Charles Darwin on the upward growth of corals with the rise of sea level. Davis's early work in **geomorphology** clarified and combined the findings of John Wesley Powell, Grove Karl Gilbert, and others. Using what was known about erosion processes, Davis developed the concept of a geographical cycle to explain how landforms, created by uplift, were eroded over long periods of time. This work led to a system of landform description based on structure, process, and time (or stage) that went well beyond the work of earlier scientists.

After retiring from Harvard in 1912, Davis tried to rescue Charles Darwin's discredited theory of coral reef formation. Building on earlier work of James D. Dana, Davis attempted to demonstrate the usefulness of geomorphological evidence in explaining how such reefs are formed.

Davis was involved in founding several professional associations, including the Geological Society of America and Association of American Geographers. He was very active in the scientific community, publishing many papers as well as two textbooks on physical geography for high school students. SEE ALSO CAREERS IN EXPLORATION AND MINING; DANA, JAMES DWIGHT; DARWIN, CHARLES ROBERT; GILBERT, GROVE KARL; LANDSCAPE EVOLUTION; POWELL, JOHN WESLEY.

Debris in Earth's Orbit

Thousands of pieces of debris currently orbit the Earth and present a challenge to the safe and economic use of space near our planet. This debris includes both natural objects, such as meteoroids and debris from comets, and man-made objects, such as nonfunctioning spacecraft and spacecraft debris.

Over 99 percent of this debris is tiny—pieces between 1 mm and 1 cm in diameter—but even tiny particles present a large problem because of the speeds at which they can travel. A 1-cm object traveling at 10 km/second (the average speed of debris in a low orbit) has the same energy as a 1,000-kg truck traveling at 60 km/hour. At these speeds, even dust-sized debris can cause serious damage to space travelers or to the surface or structure of a satellite.

Most man-made debris results from explosions, many of which are caused by flaws in spacecraft design. Fuel tanks may corrode and cause a mixing of fuels, which then explode, or a rocket's motor might fail, leading to an explosion. However, some explosions are intentional, such as those associated with weapons testing. Normal spacecraft operations also produce debris such as particles from the firing of rocket motors and even human wastes from manned spacecraft.

Several countries track space debris with radar, telescopes, and impact detectors aboard spacecraft. The U.S. Space Surveillance Network (SSN) is the main source for tracking debris, but the National Aeronautics and Space Administration (NASA) as well as organizations in the former Soviet Union, Germany, and Japan, also observe debris. Research is currently under way in the United States, Europe, Russia, and Japan to find ways to prevent the creation of new debris and to reduce existing debris. SEE ALSO COMETS; SATELLITES, ARTIFICIAL.

Desalination

Desalination is the process by which dissolved salt is removed from water. Saline water—that is, water with a high salt content—is not suitable for most human, agricultural, or industrial uses. The idea of desalination has been around for almost 2,000 years, but only in the last 50 years or so have people developed the technology to create water treatment plants for this purpose. By the early 1990s, enough desalination plants existed worldwide to treat 3.7 billion gallons of water per day. Many of those plants provide drinking water to the Middle East. In the United States and Europe, however, desalination is used mainly to treat industrial water sources.

Saline Water. Saline water contains large amounts of total dissolved solids (TDS) as well as small quantities of organic material and dissolved gases. Most of the TDS is inorganic salts, compounds made of positive metallic **ions** (calcium, sodium, ammonia) bonded to negative ions (chloride, sulfate, carbonate). These salts are formed by **erosion** of soil on or beneath the Earth's surface and by volcanic activity that releases gases from deep within the

ion atom or molecule that has a positive or negative electric charge

erosion wearing away of land by wind and water

93

Earth. These materials dissolve in water and are transported through rivers and streams to the ocean.

Types of Water.   For desalination purposes, water is separated into four categories: freshwater, brackish water, seawater, and brine. The desalination definition specifies the TDS level for freshwater as below 1,000 milligrams per liter (mg/l); for brackish water, between 1,000 and 35,000 mg/l; for seawater, above 35,000 mg/l. Brine is the concentrated salt solution that remains after pure water has been removed during desalination.

Brackish water is found in lakes, rivers, streams, and groundwater sources beneath the Earth's surface. It occurs when freshwater flows over or through mineral deposits and partially dissolves them. It is also found in such places as bays and the mouths of rivers, where freshwater mixes with seawater. Certain human activities—industrial processes, sewage disposal, and drainage from irrigated land—can produce brackish wastewater as well.

Since the oceans and seas make up 97 percent of the Earth's water supply, the vast majority of all water on the planet is seawater. Typical seawater has a TDS level of about 35,000 mg/l, although seawater composition can vary greatly. Desalting seawater is more expensive than desalting brackish water, so seawater is used for desalination only when enough brackish water is not available.

Desalination Processes.   There are three main desalination processes: distillation, reverse osmosis, and electrodialysis. The process used depends on the chemistry and physical condition of the water, how much pure water must be produced, and the type of energy source available to power the plant.

In distillation, water is boiled to produce steam, which is collected and condensed to produce salt-free water. Most distillation is done by the multi-stage flash (MSF) process. This process is based on the principle that at lower pressures, water will boil at lower temperatures. In MSF distillation, heated seawater flows into a chamber where the pressure is just low enough to boil some of the water. The hot vapor is "flashed off" and condensed on tubes that

Desalinization plant in the Canary Islands.

carry fresh, cool seawater into the system. Any water that does not evaporate moves to a second chamber with even lower pressure and the process is repeated. A large MSF plant may have 50 or more chambers.

Reverse osmosis (RO) uses pressure to force water through a special membrane that filters out the dissolved salts. Most of the energy required is used for pressurization, and no heating or evaporation is needed. Large RO plants can treat over 1 million gallons of brackish water per day. RO can also be used for seawater desalination, but for very large plants—processing 5 million gallons or more per day—distillation is more economical.

Electrodialysis (ED) works by separating dissolved minerals into electrically charged ions. ED uses two membranes: one allows only positive ions to pass through; the other, only negative ions. Saline water flows between layers of these membranes, and an electric current is passed through the water to separate the ions. This process produces alternating layers of purified water and brine. Because of the electrical energy required, ED is normally used only for brackish water with a low salt content.

Regardless of the type of process used, the brine produced must be disposed of in an environmentally acceptable manner. This is often done by pumping the brine into the ocean or some other saline water source, such as groundwater that already has a high salt content. However, pumping brine into saline groundwater may contaminate fresh groundwater at a deeper level. Where no saltwater source is close by, brine may be disposed of by evaporation, leaving a solid residue of salts. SEE ALSO FRESHWATERS OF EARTH; GROUNDWATER; SEAWATER, CHARACTERISTICS OF; VOLCANOES; WEATHERING AND EROSION.

Desertification

See Land Degradation and Desertification.

Dinosaurs

GT See Geologic Timescale on page viii.

extinct having died out completely, leaving no direct descendants (refers to a specific grouping of organisms)

paleontologist scientist who studies prehistoric plants and animals through fossils and other remains

Dinosaurs were the dominant large animals on land from their origin during the Triassic period[GT], around 235 million years (Ma) ago, until they became **extinct** at the end of the Cretaceous period, about 66 Ma ago. **Paleontologists** generally classify dinosaurs as reptiles, a group of animals represented today by lizards, snakes, crocodiles, and turtles.

There were hundreds of different kinds of dinosaurs. Some were the largest land animals of all time, weighing more than 100 tons. Others were no bigger than chickens. Some walked on four legs; others, on two. Among them were ferocious hunters, vegetarians, scavengers, and egg stealers. Some were simply covered in scaly skin, others sported bony armor, knobs, crests, spikes, or horns. Scientists have been studying dinosaurs since the early 1800s, and they continue to make new discoveries about these remarkable animals.

The "Terrible Lizards"

In 1841 Professor Richard Owen of Great Britain coined the name *Dinosauria* for a group of large fossil animals that had begun to attract attention. The term comes from the Greek words meaning "terrible lizards." Scientists would soon discover an amazing variety of these long-vanished animals.

First Discoveries. The first recorded finding of a dinosaur bone was in England in 1676, although there were probably many similar findings before then. The modern study of dinosaurs began in the early 1800s with the discovery of several fossils, again in England. By the 1840s, these ancient creatures were known to the scientific community and the public as dinosaurs. The second half of the 1800s brought a great number of discoveries of fossil dinosaurs in North America, especially from the region just east of the Rocky Mountains. Not all of the finds suggested large animals—some dinosaurs appeared to have been fairly small. Scientists realized that a window had opened onto an entire, unsuspected branch of the animal kingdom.

Classifying Dinosaurs.   Dinosaurs belong to a division of reptiles called the Archosauria, or "ruling reptiles." The only archosaurs that remain today

Dinosaur tracks at the Dinosaur Trackway site in the Painted Desert in Arizona.

crocodilians group of aquatic reptiles that includes alligators, crocodiles, and gavials

bipedal two-footed; walking on two legs

Eggs, Nests, and Parental Care

Adult dinosaur skeleton fossils have been found with baby skeletons inside their rib cages. The adults may have eaten the babies—but perhaps some species of dinosaurs gave birth to live young. Most dinosaurs, however, are thought to have laid eggs. Fossilized eggs, usually laid neatly in nests, have been found in many parts of the world, including Mongolia's Gobi Desert, southern France, Montana, and China. The number, size, and arrangement of the eggs and the type and arrangement of the nests vary widely. Scientists believe that some dinosaurs were helpless when they hatched, an idea that suggests that dinosaur parents must have fed and protected their young for a time, as do many species of birds.

carnivore meat eater

herbivore plant eater
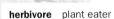

are **crocodilians,** although modern birds apparently are descendants of archosaurs. Both dinosaurs and crocodilian ancestors evolved from reptiles called thecodonts, which were similar to modern crocodiles in size and appearance but lived entirely on land.

The limbs of earlier reptiles grew straight out from the sides of their bodies. The most striking characteristic of the dinosaurs is that their limbs were positioned under their bodies, with knees pointing forward and elbows pointing backward. Dinosaurs therefore stood and walked differently from other reptiles, and some were **bipedal.**

Dinosaurs appear to have lived in all parts of the world; their fossils have been found on every continent. Fossils are most common in rocks formed in swamps, lake beds, and rivers, where the sediments covered and preserved the bodies. Large numbers of dinosaur eggs have also been found, and some of these contain fossils of unhatched dinosaurs.

Myths and Misconceptions. One popular myth assumes that all dinosaurs were large, but many, in fact, were quite small. Another misconception is that dinosaurs were slow, clumsy, stupid, and inefficient and that they became extinct for that reason. On the contrary, dinosaurs were highly successful, surviving as a group for 170 Ma. Modern scientists also know that at least some of them were limber and quick.

Another common mistake is to think that all large ancient reptiles were dinosaurs. The dinosaurs shared the world with large flying reptiles called pterosaurs (the well-known pterodactyl belonged to this group) and with large marine reptiles such as plesiosaurs and ichthyosaurs. None of these creatures were dinosaurs, although pterosaurs were fairly closely related. As far as scientists now know, there were no flying, sea-dwelling, or burrowing dinosaurs.

Types of Dinosaurs

Almost all dinosaurs fall into one of three categories: theropods, sauropodomorphs, and ornithischians. The oldest known Triassic deposits containing dinosaur fossils include fossils of theropods and sauropodomorphs, but the ornithischians did not appear until the beginning of the Jurassic period, 208 Ma ago. All three groups lived until the very end of the Cretaceous period.

Theropods. The theropods were bipedal **carnivores.** Within this group were both small, lightly built animals and large, heavily built ones. Subgroups include the "ostrich dinosaurs," the fish-eating dinosaurs, and the egg stealing dinosaurs.

Not long before dinosaurs became extinct, a group of gigantic theropods called the Tyrannosauridae appeared. Among them was the fearsome *Tyrannosaurus rex,* with enormous jaws, teeth 15 cm long, tiny forelegs, and huge, clawed hind feet.

Sauropodomorphs. Most sauropodomorphs were four-legged **herbivores.** The most familiar dinosaurs of this group are the sauropods of the Jurassic and Cretaceous periods. Sauropods had vast, elephantlike bodies; long necks; small heads; and long, whiplike tails. Among them were were the largest land animals known.

Scientists once thought that these immense creatures spent much of their lives in lakes, marshes, and swamps, where the water would help support

A painted Sun glows in a dinosaur exhibit in the Natural History Museum in Albuquerque, New Mexico.

anatomical having to do with anatomy—the structure and arrangement of body parts

quadrupedal four-footed; walking on four legs

their weight. Recent research, however, suggests that the sauropods most likely spent their entire lives on land.

Ornithischians. The ornithischians were herbivores. There were large and small ornithischians, some that walked on two legs and some that walked on four, but all shared certain distinctive **anatomical** features, particularly a pelvis similar to that of modern birds (the word *ornithischian* means "bird-hipped").

The **quadrupedal** plated and armored dinosaurs were ornithischians. So were the iguanodons, who walked on two or four legs, and the hadrosaurs, or "duck-billed" dinosaurs. The ceratopians were a group of ornithischians that resembled huge reptilian rhinoceroses. They roamed the plains in vast herds during the late Cretaceous period.

Current Controversies

Dinosaur research is a lively area of paleontology. New fossil finds and new theories continue to challenge established ideas. Three areas of particular interest to dinosaur researchers today are temperature regulation, the dinosaur-bird link, and the cause or causes of the dinosaurs' extinction.

Temperature Regulation. Until about 1970, scientists thought that dinosaurs, like all modern reptiles, were cold-blooded. The body temperature of cold-blooded animals varies according to the surrounding temperature, often requiring sunlight to maintain it at levels high enough for normal activities and functions.

Later paleontologists John Ostrom and Robert Bakker proposed that dinosaurs were warm-blooded. Warm-blooded animals, such as mammals and birds, generate their own body heat from the fuel (food) they consume. Other scientists have criticized the theory of warm-blooded dinosaurs for its lack of solid evidence. Today the most common view is that only some of the smaller, extremely active theropods may have been at least partially warm-blooded. If most dinosaurs were cold-blooded, their presence in a fossil deposit indicates that the climate must have been fairly warm when that deposit formed.

Dinosaurs and Birds.   As early as the mid-1800s, scientists noted similarities between dinosaurs and birds. In 1973 Ostrom offered considerable evidence that modern birds had **evolved** from dinosaurs. Surprisingly, most scientists agree that birds are related not to the "bird-hipped" ornithischians but to theropods. The details of that link, however, are still not clear. Birds may have developed directly from dinosaurs, or birds and dinosaurs may have evolved independently from a common ancestor. Discoveries in the 1990s of the fossils of very early birds and of feathered dinosaurs may shed light on the connection.

Extinction of Dinosaurs.   The most common question in all of paleontology concerns the disappearance of the dinosaurs. In 1993 one researcher found that 89 possible causes of dinosaur extinction had been published. Scientists do not yet know which of them, if any, is correct.

One theory that has gained much attention since the early 1980s is that the impact of an asteroid or comet produced an explosion whose airborne debris blocked out the Sun's light for years, resulting in environmental changes that caused the death of the dinosaurs and many other forms of life. We know from the geological record that many such impacts have occurred (including a large impact in Mexico's Yucatán Peninsula around the end of the Cretaceous period); still, many paleontologists feel that the impact theory does not explain why some organisms died while others survived. In addition, it is not certain that all dinosaurs died out at the same time. Further research is needed to determine how the dinosaur extinction compares with other mass extinctions. SEE ALSO Amphibians and Reptiles; Extinctions.

evolve   change over time

Drilling for Scientific Research

Since the beginning of recorded history, people have wondered what lay beneath the Earth's surface. However, only recently, with the development of deep-crust drilling, have scientists been able to answer questions about this underground environment. In the past 40 years, drilling has enabled researchers to learn more about the composition of the Earth's crust and geological processes such as earthquakes and **plate tectonics.**

plate tectonics   theory used to explain continental drift (the movement of continents over the surface of the Earth) and other geological processes

History of Scientific Drilling

Many early scientists speculated about the composition of Earth's interior, but it was not until the 1880s that serious proposals to explore the crust were put forward. In 1881 Charles Darwin expressed the need for a drill hole to produce "cores for examination from a depth of 500 to 600 feet" from the Pacific and Indian Oceans. His request was finally acted upon in 1897, but the effort achieved no concrete results. In 1922 Dr. Thomas A. Jaggar, founder of the Hawaiian Volcano Observatory, directed the drilling of four holes into the floor of Hawaii's Kilauea volcano. Instruments were lowered into the holes to record rock temperatures at depths of 7 to 24 m. This project led to a series of drill holes in the Hawaiian Islands that provided information concerning the formation of the islands and the nature of ocean crusts. Despite these successes, a long-term scientific ocean drilling program did not begin until 1960.

mantle   region of the Earth between the molten core and the outer crust

continental margin   region where continental crust meets oceanic crust

Ocean Drilling

Ocean drilling consists of drilling in ocean waters from ships and platforms and drilling on island bases. The earliest efforts were undertaken by Japanese scientists in the mid-1930s to try to determine the relationship between coral formations and the rocks upon which they grew. Later, in the 1950s, successful efforts led U.S. geoscientists to design the first major ocean drilling project. Since the Earth's crust is much thinner on the seafloor than on the continents, ocean drilling was considered the most practical way to sample rocks from the **mantle.** The goal was to reach the Mohorovičić discontinuity, or Moho—the boundary between the crust and the mantle.

In 1960 the National Science Foundation (NSF) approved funding for the program, called Project Mohole, and in 1961 a hole was drilled into the ocean floor through 3,355 m of water, north of the Hawaiian Islands. It was soon realized that a larger vessel was needed to drill down to the Moho, but the costs of such a project were too great. However, by 1965 several U.S. universities had joined together and drilled 14 holes into the Atlantic Ocean floor. The following year, the NSF funded the Deep Sea Drilling Project (DSDP), and by 1984 more than 96 km of core samples had been produced for scientists to study. The theory of seafloor spreading had been confirmed by this drilling program (then international). The DSDP, later renamed the Ocean Drilling Program, has collected data related to areas such as the composition of deep portions of oceanic crust, the nature of **continental margins,** ocean circulation, and evolutionary change in marine organisms.

Continental Drilling

Because ocean crust is much younger than the continental crust, examining it yields information about only a small part of Earth's history—the most recent 200 million years. To learn more about Earth's history requires drilling into

Scientists from the Coast Guard icebreaker *Polar Star* drill holes into the pack ice in the Arctic Ocean off Nome, Alaska, in 1970.

The *Glomar Challenger,* a customized drill ship for deep sampling of ocean crust.

the far thicker continental crust. Since the oil and gas industry had drilled holes as deep as 9,455 m, scientists felt that continental drilling to great depths was possible. A drilling program began formally in the United States in 1985. The most productive drilling produced a hole 3,512 m deep near California's San Andreas Fault, a site of frequent earthquakes. This particular project gathered data regarding the nature of the fault, including pressure and temperature relationships, physical characteristics of the rock, and other information that could contribute to more reliable earthquake predictions.

Continental drilling has been and continues to be a worldwide activity; the oldest such program began in the former Soviet Union. The world's deepest hole was drilled in northwest Russia's Kola Peninsula and attained a depth of 12,261 m. In addition to data about the geology of this area of the Earth, the Russian project has led to new advances in drilling technology. Germany, Belgium, Canada, France, Switzerland, Sweden, Iceland, Japan, New Zealand, and Great Britain have also conducted continental drilling for a variety of purposes. These include finding mineral and fuel reserves, locating safe places underground to store nuclear waste, evaluating geothermal energy resources, and advancing geologic knowledge. As questions about the nature of the planet continue to be raised, drilling programs will undoubtedly become more ambitious in an effort to drill ever deeper and unlock the mysteries beneath Earth's surface. SEE ALSO CONTINENTAL MARGINS; EARTH, STRUCTURE AND COMPOSITION OF; EARTHQUAKES; GEOLOGIC TIME; GEOTHERMAL ENERGY; HAZARDOUS WASTE DISPOSAL; LIFE, EVOLUTION OF; MANTLE; PLATE TECTONICS.

Du Toit, Alexander Logie

1878–1948
South African geologist

Highly regarded in South Africa for the extent and quality of his geological mapping, Alexander Logie du Toit was responsible for much of the geological understanding of the southern third of Africa. In the world outside Africa, du Toit is best known for his support of the **continental drift** theory, an idea first proposed by German scientist Alfred Wegener.

continental drift   theory that the continents move over the surface of the Earth on large segments of Earth's crust and uppermost mantle (region just below the crust)

sedimentary rock   rock formed from deposits of sediments (soils, rock particles, and other materials) over long periods of time

igneous rock   rock formed when molten rock, which is produced under intense heat, cools

plate tectonics   scientific theory used to explain continental drift (the movement of continents over the surface of the Earth) and other geological processes

Born in Rondebosch, South Africa, du Toit attended college in nearby Cape Town and then went to Scotland to study mining engineering. In 1900 and 1901, he studied geology at the Royal College of Science in London. Du Toit returned to South Africa in 1903 and became assistant geologist for the Geological Commission of the Cape of Good Hope. He remained there until 1921, when he was appointed chief geologist to the Department of Irrigation.

Between 1903 and 1921, du Toit carried out the geological mapping for which he became famous. He became a noted authority on the Karroo System—a system of **sedimentary** and **igneous rocks** that are widely distributed throughout southern Africa and extend to other continents as well. This work led him to embrace the theory of continental drift, which brought him world fame.

After 1921 du Toit spent much time gathering evidence to support the continental drift hypothesis. After retiring from government service in 1927, he worked for mining companies and traveled widely, still pursuing evidence of continental drift. His best-known work, *Our Wandering Continents* (1937), presented the most eloquent and closely argued work in support of continental drift prior to the development of **plate tectonics** theory in the 1950s and 1960s. SEE ALSO IGNEOUS ROCKS; PLATE TECTONICS; SEDIMENTS AND SEDIMENTARY ROCKS; WEGENER, ALFRED LOTHAR.

Earth, Motions of

axis   line through the center of the Earth about which the Earth rotates

Although an observer feels no motion as Earth travels through space, the ground beneath the observer's feet moves swiftly along a complex path. This path can best be understood by looking separately at each contributing movement.

Rotation.   Early watchers of the skies observed that the Sun and stars seem to move steadily from east to west in a giant circle. The Moon and planets, which make their own obvious motions in the sky, are also carried through the same circle. The reason for this appearance of steady motion is Earth's rotation. As the Earth turns on its **axis,** or rotates, the sky seems to move in a direction opposite that of Earth's rotation. The passage of one rotation is the traditional basis of time, a day of 24 hours that was also originally used to define the second as exactly $1/86,400$ of one rotation.

The rotation of Earth has many small variations, however. Modern clocks, based on quartz crystals and atomic vibrations, show that day length changes periodically and is gradually lengthening. Interactions between the fluid and solid parts of the planet are the main cause of these variations. For example, the tides cause Earth's rotation to slow because the bulge in the ocean waters caused by the pull of the Moon's gravity takes time to form and also to sink back down. Earth rotates faster than the Moon passes overhead. As a result, the bulge is always a small distance ahead of the position of the Moon: high tide occurs shortly before the Moon has passed. The Moon, by attracting the tidal bulge, slows down Earth's rotation. Because the official second used by scientists is based on a very precise atomic clock, it is necessary to add 1 second to a selected day about once a year to account for variations in Earth's rotation.

Revolution.   Along with the daily movements of the Sun and stars, the ancient watchers of the sky saw that the path of the Sun through the sky appears to move slowly back and forth. Regular changes in weather accompany this motion, which is to the north and south. In the northern hemisphere, the Sun reaches its most northern point at the start of summer, for example.

This apparent motion of the heavens is also the result of the motion of Earth. Over the course of a bit more than 365 rotations, Earth travels once around the Sun in a nearly circular path called an ellipse. The path is Earth's orbit about the Sun. The Moon similarly completes one orbit around the Earth in about 29 days. The Moon is inclined some 5 degrees from Earth's orbit and pulls Earth slightly from side to side, out of a purely elliptical path.

Each passage of the Earth around the Sun is called a revolution. The time of one revolution is called a year. The main motions of Earth combine its rotation around its axis with its revolution around the Sun. Since one revolution is 365.256 rotations, a year is 365.256 days. Calendars try to keep the year in line with the seasons by adding a day, called a leap day, every 4 years to make 366 days instead of 365, but even then the match is not quite exact, and the leap day is omitted every 400 years unless the year ends in three zeros.

The seasonal change in weather each year is not caused by the slight changes in distance to the Sun that result from the elliptical orbit. Earth is tilted by about 23.5 degrees from a line perpendicular to the plane in which it orbits. The rotation of Earth maintains this tilt, just as the rotation of a spinning top keeps it upright. The tilt positions the northern hemisphere toward the Sun for half of the orbit and the southern hemisphere toward the Sun for the other half. This shifting relationship is the cause of the changing seasons and also of the north-south change in the apparent path of the Sun.

Wobbles. The axis of a spinning top, once it starts to lean, begins to move in a small circle, a shift in motion called precession. Precession also happens to Earth, and for the same reasons. The change in tilt of Earth's axis is slow—about 26,000 years to make one circle—but noticeable. It was discovered by Babylonian astronomers and known to the ancient Greeks.

There are two much smaller wobbles in Earth's axis, one discovered in 1891 and the other first noticed in the 1980s. The axis also has been drifting slightly, possibly as a result of the changing shape of Earth caused by large ice caps forming or melting over geologic time.

Bounces. If all of these motions were not enough, Earth also moves back and forth slightly as a result of large earthquakes. The impact of a large asteroid or comet, such as the one that struck Earth about 66 million years ago, also sets off such bouncing.

All of these motions are related primarily to the Sun. In addition the whole solar system, including Earth, is always moving through space—first, in a giant rotation of the Milky Way galaxy and second, as a part of the galaxy's movement in a particular direction through the universe. SEE ALSO EARTH, STRUCTURE AND COMPOSITION OF; TIDES.

Earth, Origin of

The origin of Earth began with the formation of the solar system, some 4.55 billion years ago. The solar system had its beginnings in a molecular cloud of gas and dust called the solar nebula.

The Solar Nebula. During the early stages of the solar nebula's development, solid material called interstellar dust was concentrated at the middle of the nebula. Some of the material, composed of metal and silicate dust, clumped together to form millimeter-sized dust balls. This material was heated, although the exact way in which it was heated is still unknown. Lightning, chemical energy, frictional heat, and heat from magnetic processes

Earth, Origin of

condensation   change from a gas to a liquid

meteorite   solid body that has fallen to Earth from space

density   amount of mass (matter that causes an object to have weight) in a unit of volume

volatile   easily vaporized at moderate temperatures and pressures

refractory   able to withstand very high temperatures without melting or vaporizing

mantle   region of the Earth between the molten core and the outer crust

plate tectonics   theory used to explain continental drift (the movement of continents over the surface of the Earth) and other geological processes

have all been suggested as possible causes. The material also underwent other processes, such as evaporation and **condensation.** Heating caused the clumps of material to melt, producing round objects called chondrules, which are found in **meteorites** known as chondrites.

The chondrules eventually accreted (clumped together), due to impact with one another, into asteroid-sized objects called planetesimals. As more material accreted, the planetesimals gradually grew into the planets that make up the solar system. The later stages of this process were marked by the accretion of very large objects with existing planetesimals. This accounts for the fact that every planet is tilted on its axis (Earth is tilted by 23.5°). Only a collision with a large object would cause such a tilt, since large numbers of smaller objects accreting over a period of time would tend to cancel each other out.

Earth-Moon Origins.   Although several planets have satellites, the Earth-Moon system is unique in several ways that are probably related to how the system was formed. For example, the Moon has a low **density** compared to Earth, so it is likely that it contains much less iron than Earth. Also, the Moon contains no water and very small amounts of **volatile** elements such as potassium and lead. However, it probably has unusually large amounts of **refractory** elements such as calcium, aluminum, titanium, and uranium. Despite these differences, the composition of the Earth's **mantle** and that of the Moon do show similarities.

Several theories have been proposed to explain the origin of the Moon. However, the most accepted hypothesis is that the Moon was created when a large object collided with the Earth, causing material to be ejected from Earth's mantle. The material was then captured in Earth's orbit and eventually accreted to form the Moon. According to this hypothesis, the Earth's iron core had already formed by this time, so most of the Earth's iron had been removed from its mantle. This idea, if true, would explain why the composition of the Moon is similar to that of the Earth's mantle; it also accounts for the low amount of iron in the Moon. The Moon's low content of volatile elements could be explained by their vaporization from the mantle material as it blasted into Earth's orbit. Metal that was carried away from Earth by the impact may have formed a very small lunar metallic core of iron and nickel. There is also evidence that melting of the outer portion of the Moon produced a magma ocean that caused additional chemical changes.

The Early Earth.   The accretion of Earth lasted from 50 to 200 million years after the solar system first formed. However, there are no existing rocks that date back to the first half billion years of Earth's history. Therefore, knowledge of the Earth's earliest history can only be gained by studying later rocks and meteorites to determine how the Earth developed into its present state. Large objects were certainly striking Earth during this time, and these impacts probably caused extensive melting of early Earth, producing a magma ocean like that on the Moon. Unlike the Moon's, the Earth's magma ocean did not leave a clear chemical record in the upper mantle, possibly because of Earth's larger size and more intense geologic processes such as **plate tectonics.** These geologic processes may have mixed the solidified material that had made up the magma ocean back into the mantle later on. In any case, the structure of the outer portion of Earth above a depth of 670 km has been interpreted as the remains of a magma ocean.

The Earth's core probably formed at an early stage of accretion, when meteorites that contained significant amounts of iron hit the young Earth. Over time iron and nickel metal separated from the silicate (rocky) portion of the hot

Earth and sank to the center of the planet to form the core. This process also stripped the crust and mantle of elements that are chemically attracted to iron and iron sulfide, such as cobalt, nickel, copper, gold, and platinum. Several models have been proposed to explain how this happened.

Equilibrium models suggest that the depletion of metal in the mantle represents a state of chemical equilibrium, or stability, between metal and mantle silicates. Disequilibrium models approach the problem differently. One such model says that the Earth's core formed first, removing most metal from the mantle. Later on, material containing metallic elements, possibly from the solar nebula or even from the metal cores of large planetesimals, accreted onto the Earth, adding some metallic material to the mantle. Another disequilibrium model says that silicate material that had been stripped of metallic elements was mixed into the mantle, diluting the amounts of metallic elements in the mantle as a whole. Any of these models may be correct, and scientists are studying processes at the boundary between the core and mantle in hopes of finding additional clues to the chemical state of Earth's core.

Earth's atmosphere is very poor in the "noble," or inert, gases (helium, neon, argon, krypton, and xenon) compared to the solar system as a whole. This may be because a strong **solar wind** scattered the gases before the young Earth became large enough to capture and retain a primitive atmosphere. It also may be that the early atmosphere was blown off by the impact of large objects. The mixture of **isotopes** of some of the noble gases in Earth's atmosphere indicates that the atmosphere was created as heat drove gases from the mantle during the first half billion years after Earth's accretion. However, free oxygen did not appear in the atmosphere until oxygen-producing bacteria became abundant, from 2.9 to 2.5 billion years ago.

The origins of the Earth's oceans are still uncertain. Since metal and liquid water cannot exist in equilibrium as can metal and mantle silicates, the oceans must have formed after the core. However, whether water accreted from ice-rich comets or from planetesimals that contained volatile elements is not known. The early history of the continental crust is also unknown, since most of the existing continental crust was formed 2 to 2.5 billion years ago, and even the oldest continental rocks date from half a billion years after the formation of Earth. There is evidence that some continental crust existed on the earliest Earth, but it was unlikely to have been similar to modern crust. Some scientists suggest that an early basaltic crust was present but was later mixed back into the mantle. SEE ALSO ATMOSPHERE, EARTH'S; EARTH-MOON SYSTEM, EARLIEST HISTORY; EARTH'S CRUST, HISTORY OF; MOON; OCEANS, EVOLUTION AND STRUCTURE OF; SOLAR SYSTEM.

solar wind flow of charged atomic particles, mostly hydrogen (H+), away from the outer layers of the Sun

isotope one of two or more forms of the same element; isotopes differ in mass (the amount of matter that causes a body to have weight in an area influenced by gravity)

Earth, Physics of

Much current knowledge about the structure, properties, and behavior of the Earth has come from the science of physics. The term *physics of the Earth* commonly refers to the use of physics to answer only basic theoretical questions, whereas the term *geophysics* also includes related sciences, such as seismology, and certain techniques used in the search for mineral deposits and oil and gas.

density amount of mass (matter that causes an object to have weight) in a unit of volume

mass amount of matter that causes an object to have weight when it is in an area influenced by gravity; commonly measured in kilograms

Density of the Earth. The **density** of the entire Earth can be determined by applying the law of gravity. Logically, this law reveals that the Earth as a whole is approximately twice as dense as the rocks in its crust. Therefore, the interior (or some part of it) must be much denser than the crust; increased pressure at the Earth's center explains only a small part of the difference. The degree to which the Earth's **mass** is concentrated at the center determines its rotational inertia—that is, how much force is required to get it to rotate. Scientists have measured the rotational inertia by observing the gravitational effects of the Moon and Sun on the orientation of Earth's axis of rotation. These measurements were used in the 1930s to create detailed models of Earth's density structure.

However, the most precise means of measuring the density of the various layers of Earth's interior comes from seismology, the study of earthquakes and of the associated waves that travel through the Earth. The speed of these "seismic waves" can be measured by timing how long it takes them to reach recording stations around the globe. Wave speeds generally increase with pressure, and pressure increases with depth, so if the Earth's interior were uniform in composition, the speed of seismic waves would increase uniformly according to depth. This is approximately true in the lower **mantle**—below 670 km deep—but not in the upper mantle, where seismic wave speeds change suddenly at particular depths. This fact suggests that the upper mantle contains several layers that vary not only in pressure but also in composition.

Calculating the exact densities of these layers is possible by observing standing waves—seismic waves that continue to travel back and forth throughout the Earth for several days. The frequency of these vibrations depends upon the internal structure of the Earth, so by studying them, scientists have been able to create a profile of density at different depths. By comparing the densities of various Earth minerals and rocks with this profile, they can largely determine the composition of the various layers of the mantle.

Earth's Interior Structure. One theory that can explain the layers of differing densities in the Earth's mantle is that the mantle is composed of various forms of the mineral olivine. As pressure increases with depth, olivine undergoes a series of transformations to denser **crystal** structures. The difference between the upper and lower mantle may be due to the different crystal structures of olivine at greater depths. However, other scientists feel that this theory may not adequately explain the properties of the mantle.

If the Earth's mantle is indeed composed of olivine, its core must have quite a different composition. No combination of the elements that make up olivine

mantle region of the Earth between the molten core and the outer crust

crystal any solid whose internal structure is arranged in a repeating, orderly pattern

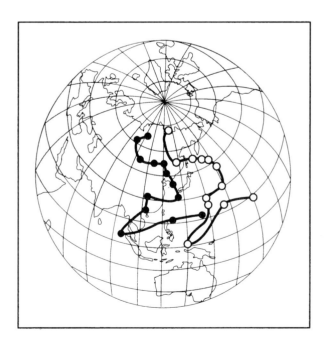

Pole paths for Europe and North America between 500 and 200 million years ago. These paths would coincide if the continents were moved to close up the Atlantic Ocean.

(magnesium, iron, silicon, and oxygen) could have the density that must logically occur at the core. The presence of large quantities of iron in meteorites, the Sun's atmosphere, and the other planets of the solar system, points to metallic iron as the main component of Earth's core. This is a reasonable guess, since the core must be metallic in order to carry the electric currents that are responsible for the Earth's magnetic field. However, the outer core is not sufficiently dense to be composed entirely of iron. It is therefore identified as liquid iron with some lighter elements dissolved in it, probably oxygen and possibly sulfur, silicon, carbon, and hydrogen as well. The solid inner core is mostly pure iron with smaller amounts of nickel and other elements.

Until the core was discovered by seismologists around 1900, no one could satisfactorily explain Earth's magnetic field. Early students of Earth's magnetism suggested that it resulted from **extraterrestrial** factors, since geomagnetic storms seemed to be related to auroras and sunspots. Even though Carl Friedrich Gauss proved the internal origin of the field in the 1830s, he did not know what caused it. Then, in 1919, Joseph Larmor proposed that internal motions of a fluid, metallic-conducting core, when combined with rotation, could create and sustain a magnetic field.

The energy needed to drive the internal motion of the fluid must be provided by **convection.** The core is slowly cooling, causing the solid inner core to expand as the liquid outer core solidifies. During cooling, the lighter elements are pushed out of the solidifying iron and into the surrounding liquid. As a result, the layer of fluid nearest to the inner core becomes much less dense than the rest of the outer core. As the resulting lightweight fluid mixes into the outer core, it produces an up-and-down movement in the liquid iron. This process is known as compositional convection. At the same time, the lighter inner layer rotates faster than the outer layer. The combination of rotation and convection provides the motion needed for the magnetic field to sustain itself and is widely believed to play an important role in **plate tectonics.**

Magnetic Field Behavior.   Earth's magnetic field was studied long before it was explained, and early observers noted that it resembled a dipole, a magnet with two poles of opposite charge. About 80 percent of the Earth's magnetic field can be accounted for by a dipole field tilted at about 11° to the Earth's north-south axis. However, when the effects of this dipole field are canceled out, another magnetic field remains—a random nondipole field that grows, decays, and drifts about the globe. The dipole field also varies in strength and direction over time, but more slowly. In fact, the total magnetic field has reversed itself repeatedly, with the magnetic North Pole replacing the South Pole, and vice versa.

Since many rocks retain the magnetization they acquired when they were formed or last heated, examining them to determine their ages and magnetic properties allows scientists to trace the direction of the Earth's magnetic field through time. This technique, known as paleomagnetism, reveals that the Earth's magnetic pole has wandered relative to Earth's axis of rotation. The position of the pole where it intersects Earth's surface follows a path across the globe that varies from one continent to another. However, if the continents were moved in such a way as to close up the space now occupied by the Atlantic Ocean, the paths would match rather well. This finding, first demonstrated in the 1950s, indicates that the continents themselves have drifted over time and laid the foundation for the theory of plate tectonics.

Other Studies.   Physics can be used to study the Earth in a number of other ways as well. For example, geodesy is a branch of physics that is linked to

---

extraterrestrial   from beyond the Earth

convection   circulatory movement in an unevenly heated fluid (liquid or gas); cooler material is denser and sinks in an area influenced by gravity, while warmer material is less dense and rises

plate tectonics   theory used to explain continental drift (the movement of continents over the surface of the Earth) and other geological processes

tectonic plate large segment of Earth's crust and uppermost mantle (region just below the crust) that moves as a unit over the Earth's surface, floating on a partially molten layer of rock below

mass amount of matter that causes an object to have weight when it is in an area influenced by gravity; commonly measured in kilograms

sediment soils, rock particles, and other materials that are deposited over time and make up the ground, whether on dry land or at the bottom of a body of water

density amount of mass (matter that causes an object to have weight) in a unit of volume

seismic waves vibrations associated with earthquakes, explosion, and impacts of solid bodies that fall to the Earth from space; these vibrations travel through the Earth

astronomy and deals with the Earth's shape, variations in its gravity, and its motion in space. Physics has also been used to investigate the forces that drive plate tectonics. Comparing the energy produced by mantle convection with that released during plate subduction (sinking of plates into the upper mantle) can help answer questions about what ultimately happens to **tectonic plates** after they sink. SEE ALSO CONTINENTS, EVOLUTION OF; EARTH, STRUCTURE AND COMPOSITION OF; HEAT BUDGET OF THE EARTH; MAGNETIC FIELD, EARTH'S; PLATE TECTONICS.

Earth, Structure and Composition of

Earth has an internal structure that consists of five main layers with distinct chemical compositions and physical properties. Only the very outermost surface of Earth, including the oceans and atmosphere, is readily available for study. Information about the internal layers must be inferred, or worked out from indirect evidence. The observable portion of Earth represents an insignificant fraction of the Earth's total **mass**—less than 0.5 percent. Almost all of the mass that is available for direct study is crust, the rocky layers that form Earth's surface.

The Crust. Earth's crust is separated into two main parts. The continents, formed of less dense continental crust, rise above the oceans; the denser oceanic crust forms the bottom of the seas. Continents are largely covered with a thin layer of soil or sand, but there is continuous rock wherever one probes far enough below the loose surface layer. This underlying material is called bedrock. Although the upper layers of bedrock are often formed from **sediments** that have hardened (sedimentary rock), the lower layers of bedrock are all rock that was melted at one time or another (igneous rock). Both kinds of rock are sometimes altered in form by heat and pressure (metamorphic rock). Some of this rock is very old. The oldest rock fragments on the continents are 4 billion years old.

Study of the rock layers around the Earth reveals that the bedrock beneath the oceans is different from that beneath the continents. The oceanic crust is formed from dark, heavy rocks called basalts, but the continents are made up mostly of granitic (granitelike) rocks, which are richer than the ocean basalts in the light elements silicon, aluminum, sodium, and potassium. The continental crust also contains many other kinds of igneous, metamorphic, and sedimentary rocks. When all of these different kinds of rock are included, along with the sedimentary and sedimentary-metamorphic components of the oceans, the average **density** of the crust comes to about 2.8 g/cm³ (grams per cubic centimeter). This density is about the same as that of many familiar rocks (and also that of ordinary concrete). The density varies from one kind of rock to another, however, and denser rock tends to sink below lighter rock. Lower layers of the crust have a density of about 2.9 g/cm³ or more, whereas the top layers are only about 2.6 g/cm³ in density.

When scientists learned to use the **seismic waves** associated with earthquakes to probe below Earth's surface, they discovered that there is a change in wave velocity, or speed, that occurs with depth. Seismic waves tend to travel faster in denser materials, so a sharp increase in wave velocity indicates a sudden increase in density. Such a change in density marks the bottom of the crust. This seismic/density shift is called the Mohorovičić discontinuity, or Moho. On continents, the average thickness of the crust above the Moho is 35 km, but the crust is thicker under mountains. The greatest crustal thickness on Earth is 80 km beneath the Himalaya and the Tibetan Plateau. The ocean crust is much thinner than the continental crust. Ocean crust has an average

The variation in density as a function of radius (top panel) and depth (bottom panel).

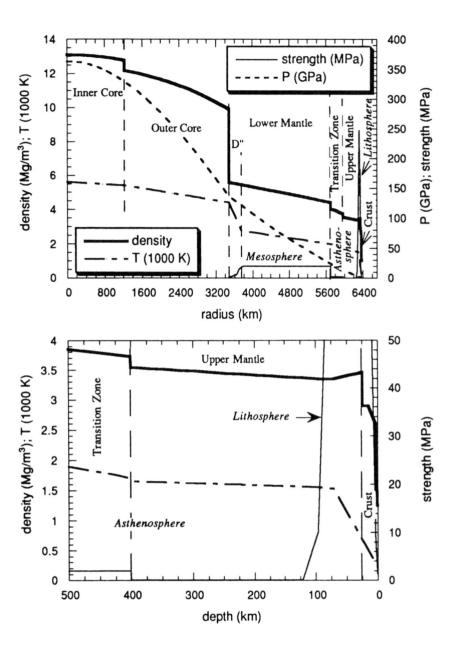

thickness of only 7 km. The average thickness of Earth's crust, when both oceanic and continental crust are included, is 16 km.

Crustal rocks are lighter than the materials below the crust. This fact is evidence that the Earth was once molten and that the minerals that form the crust separated from the heavier minerals as the molten material cooled. When the lighter materials solidified at the top, the original crust was formed. However, Earth's crust is not a continuous covering of the sphere. Instead, the crust and the material just below the crust (known together as the lithosphere) is broken into pieces called plates—a dozen or so large, rigid pieces and some smaller ones. These plates cover the whole sphere; they also move slowly with respect to each other.

The motions of the plates are connected to the production and elimination of oceanic crust. The basaltic ocean crust is forming continually at rifts between the plates, long cracks in the crust that are also called spreading centers (since the plates move away from these rifts). Furthermore, oceanic crust is

drawn down, or subducted, into the Earth at deep trenches called subduction zones at the far edges of these plates. As a result of this recycling process, nearly all of the crust of the oceans is younger than about 150 million years.

The continental crust, because of its greater thickness and lower density, does not get drawn down into the subduction zones. This difference explains why the oldest rock is all continental.

The Hydrosphere. The oceanic crust covers nearly 70 percent of Earth's surface and, as its name suggests, is covered almost completely by oceans. Water also covers large regions of the continents, either as freshwater in lakes, rivers, and ponds, or as frozen water in the form of ice caps and glaciers. This covering of water, along with water contained in reservoirs underground, is called the hydrosphere. If the hydrosphere were spread uniformly over the whole planet, it would have a depth of 3 km—slightly less than half the thickness of the crust beneath the oceans, but still a substantial part of the planet. The water from which the hydrosphere is formed is composed of the elements hydrogen and oxygen, which are relatively light, and many dissolved minerals such as salts.

The main part of the hydrosphere is the oceans. Like the plates of the crust, the parts of the hydrosphere continually move and reshape themselves. One way in which this happens is that water changes from a fresh state to a salty one. The rivers that flow across the continents start as freshwater from rain, but they dissolve salts and other minerals from the rocks and soil as they flow. These dissolved salts are carried to the ocean, where they accumulate. The saltiness of the oceans is increased by evaporation, since the dissolved minerals stay behind when water vapor enters the atmosphere. Differences in the amount of salt from one place to another and differences in temperature make some parts of the ocean more dense than others. The result is ocean currents that carry the denser water to the deeper regions, pushing the lighter water up and out of the way. The movements of such currents in oceans are thought to resemble the situation deeper in the Earth, where rocks are hot enough or under enough pressure to flow in the same way that water does—although rather more slowly.

The Atmosphere. Major divisions in the Earth's structure are closely related to differences in density. The hydrosphere, with a density close to that of pure water—1 g/cm³—lies atop the crust, which has a density of about 2.8 g/cm³. Gases are even less dense than water and form a layer above both the crust and the hydrosphere. This layer, the atmosphere, is the only part of the Earth that consists of free gases. Gases that can be observed coming from volcanoes or other crustal sources were dissolved in liquids or solids before their release. After their release they become part of the atmosphere. Water that enters a gas phase because of evaporation is considered to be part of the atmosphere rather than a component of the hydrosphere as long as it remains a gas.

Gases that are not held by some sort of container usually spread in all directions until they are so thin—of such low density—that they seem to vanish. The "container" for the atmosphere is Earth's **gravitational field,** which holds the gases close to the Earth. Because gases are all highly compressible, the lower layers of the atmosphere are much denser than upper layers. The top part of the atmosphere is so thin that it trails off into the vacuum of interplanetary space. One way to look at this sharp reduction in density of the atmosphere with height is to note that the mass of the lower 5 km is equal to the mass of all of the atmosphere above 5 km, even though the atmosphere extends hundreds of kilometers into space.

gravitational field area influenced by gravity

The main gases in the lower atmosphere are nitrogen (about 78 percent), oxygen (about 21 percent), and argon (about 1 percent). Lighter gases, such as helium and hydrogen, are found primarily in the upper levels.

The structure of the atmosphere can be described as four layers, which are distinguished by changing temperatures. The troposphere, on average the lowest 10 km of air, is characterized by temperatures that fall relatively quickly with elevation. At the top of the troposphere, or tropopause, the temperature stops falling. The layer of air above the tropopause—known as the stratosphere—is rich in ozone, a special form of oxygen. Because ozone absorbs ultraviolet radiation from the Sun, stratospheric temperatures start rising at about 20 km and continue to rise until about 47 km—at the stratopause. Earth's weather primarily develops in and affects the troposphere, but it is somewhat influenced by events in the stratosphere. The third atmospheric layer, the mesosphere, is characterized by falling temperatures from about 52 km to about 82 km—at the mesopause. In the thermosphere, above the mesosphere, temperatures again begin to rise at about 94 km. This temperature increase is due to the absorption of very short wave radiation from the Sun.

Tools for Seeing into the Earth. The crust, hydrosphere, and atmosphere can be directly observed, although the lowest layers of the crust are still hidden. All of the Earth below the crust—and that is most of the planet—must be understood from indirect evidence. To learn what lies beneath the crust or even beneath the top layers of the crust, scientists have found a variety of indirect ways to observe Earth's interior.

There are several ways to study the materials within the Earth. Earthquakes are associated with waves that travel through great distances inside the Earth before emerging at the surface. Information about the materials through which these waves pass on their journey can be obtained by studying the waves. Volcanoes bring up material from below the surface, although many scientists believe that much of the lava and gases expelled from volcanoes does not originate very far below the surface. The exact shape of the Earth, which can be calculated from gravity measurements and from satellite studies, reveals much about the strength of the materials within the Earth.

Various approaches contribute to an understanding of Earth's mass, density, and magnetism. Gravity measurements can be used to "weigh the Earth"—that is, to determine the mass of Earth, from which the average density can be calculated. Earth's **magnetic field** also can be used to infer movements and composition of the magnetic materials that create the field. The laws of motion can be used to determine how the mass within the Earth is distributed.

magnetic field region surrounding a magnet or an object carrying an electric current that affects the behavior of other magnets, magnetic materials, and electric currents

Nature of the Mantle. The layer directly under the crust is known as the mantle. *Mantle* may seem an odd term for this large, inner part of Earth, since the word means "covering." This term was introduced because earth scientists first used seismic waves to establish that the innermost part of Earth is a very dense core. The rest of Earth, covering the core, was called the mantle. A few years after the core was discovered, geologists learned from seismic waves that the crust is physically different from the layer just beneath it. Although the crust is the uppermost covering, the word *mantle* has become the standard name for the layer below the crust.

The mantle is a huge region, stretching from the lower boundary of the thin crust to a depth that is 3,480 km from Earth's center. The average thickness of the mantle is 2,867 km, and because it completely surrounds the core, it accounts for approximately five-sixths of the volume of Earth.

Earth's Meteorite Relatives

Meteorites are chunks of material from the rest of the solar system that get caught in Earth's gravitational field and fall to the Earth's surface. Scientists think that the various types of meteorites resemble the kinds of material from which Earth's core, mantle, and crust were formed. Meteorites may also have been responsible for creating these structures on Earth. Scientists think that the Earth was solid when it first began to form, but that after it reached a substantial size, large meteorites struck it with such intensity that they partially or completely melted it. Iron alloyed with other elements sank to form the core, while lighter elements rose to form the crust, producing the Earth's layered structure.

A few samples of mantle material have reached the surface because of movements within the upper mantle and crust. These rocks are high in magnesium and iron but low in silicon and oxygen, a combination that makes the rock heavy and dark, almost black. Geologists call such rocks ultramafic (where *mafic* is formed from the name of the element magnesium combined with *ferric,* a word used to describe iron compounds).

Although the mantle is considered a single layer, it—like the crust, hydrosphere, and atmosphere—consists of several different parts with different properties. Factors that affect mantle properties include pressure, temperature, and composition.

Pressure from the layers of crust and the higher parts of the mantle is very high in the Earth's interior and increases with depth. Increased pressure on a material makes the material more dense, in part simply by squeezing the atoms closer together. The Earth is also hot inside, and temperatures generally rise with depth. The effects of temperature on density, however, are opposite those of pressure. Atoms move farther apart at higher temperatures.

When the Earth was completely or partially molten, minerals and rocks composed of heavier elements tended to sink to the lower layers. Therefore, the main minerals in the mantle are mafic rocks, rich in such heavy elements as magnesium and iron.

Areas of the Mantle. Various combinations of pressure and temperature, along with the structure and composition of minerals, can affect the strength of rocky material in the mantle and how it can be deformed. At very high pressures and temperatures, minerals actually change structure. Their atoms move from one configuration to another that brings them closer together under the influence of very high pressures, or farther apart as a result of very high temperatures. Consequently, the structures of minerals in the mantle differ from the structures of minerals in the crust, with the tightest structures found deep in the mantle.

Even the very top part of the mantle is dense and hot, with a density of 3.42 g/cm^3 at temperatures below 750 K (750 kelvin, or about 477°C). In this uppermost area—the mantle component of the lithosphere and the plates into which it is broken—the material is rigid. Under high enough stresses, these rocks can suddenly break and move, causing earthquakes. At lower levels, the materials are under such high pressure and the temperature is so high that they deform more easily. Earthquakes, which require brittle fracture of rock, do not commonly occur under such conditions.

Near the top of the mantle, the interaction of pressure and temperature affects both the density and the physical state of the rock. Between about 75 and 100 km from the surface of the Earth, on average the temperature is high enough for partial melting. This semiliquid portion of the mantle, known as the asthenosphere, provides a way for the crustal plates to move over the surface of the mantle. This material can flow like the children's toy Silly Putty, slowly changing shape or even moving from place to place.

About 400 km below the surface, the first of the rearrangements of mineral structure occurs. Above 400 km, the mantle is composed primarily of a mineral called olivine. Below that depth, the chemical components of the rock are rearranged into a series of denser mineral structures known as garnet, pyroxene, and spinel. At 660 km below the surface, there is an especially dramatic change in structure to a form called perovskite. By the perovskite stage, the density has climbed to 4.38 g/cm^3. The mantle from just beneath the crust to this depth is known as the upper mantle; the boundary where the structure rearranges to perovskite is called the transition zone.

Below the transition zone, the lower mantle continues to increase in temperature and pressure. At the top of the lower mantle, the temperature is thought to be between 1600 K and 2400 K; the pressure is 240 thousand atmospheres (1 atmosphere is the pressure of the atmosphere at sea level). The increasing pressure produces higher density as well, with the density averaging 4.38 g/cm^3 at the top of the lower mantle and 5.57 g/cm^3 at the base. Temperatures vary, not just with depth but laterally (from side to side), in the lower mantle.

Near the base of the mantle, where it meets the top of the core, is a zone several hundred kilometers thick that is not yet clearly understood. The zone is nearly 3000 km away from the nearest measuring equipment; the space between the surface and the zone is filled mostly with solid rock; and the chemical interactions between the core and the mantle are not well known. Although it is possible to produce very high pressures and temperatures in small samples of rock in laboratories, it has not been possible to duplicate conditions at the mantle-core interface. **Seismologists** who have studied this region call it the D" layer, for its effects on seismic waves as they pass through it. These changes in seismic waves indicate large variations in structure throughout the D" layer.

seismologist scientist who studies earthquakes

The Core. Earth's core was the first major internal structure to be recognized by geologists. Its existence was suspected in the 1800s, and seismological evidence confirmed as early as 1906 that there is a large liquid region below the solid part of Earth. The evidence also demonstrates that the molten region is metallic, mostly iron but combined with nickel and containing some lighter elements, such as sulfur, oxygen, and perhaps carbon and hydrogen. No one knows for sure which light elements are in the core, but the mass of the core is lower than it would be if the only elements present were iron and nickel. Although the **alloy** is lighter than iron, it is suspected that there are some heavy elements in the alloy as well because nickel, cobalt, and traces of iridium and other metals are found in meteorites that are thought to be similar in composition to Earth's core. Also, heavy elements and elements that tend to associate with iron would naturally have fallen toward Earth's center when Earth was molten.

alloy substance composed of two or more metals, or of a metal combined with a nonmetallic substance

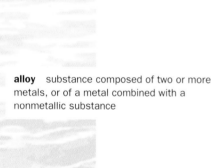

The iron of the outer core is molten because of the high temperature, estimated to be between 3,800 and 5,000 K. Even at the great pressures this deep within the planet—about 1.35 million atmospheres at the core-mantle boundary—the iron alloy at that high a temperature flows as freely as liquid water.

However, pressure and temperature continue to increase toward Earth's center. By a distance of 1,222 km from the center of the Earth, the effects of pressure overcome the effects of temperature. The iron alloy begins to solidify, or freeze. Furthermore, the entire planet, including the core, is cooling over time. As a result, the inner core is growing larger.

As the inner core freezes, the light elements in the iron tend to be forced out of the alloy, so the solid, inner core is practically pure iron. The lighter elements that are still liquid when the iron solidifies reduce the density of the liquid near the inner core. Because this liquid is lighter than the alloy above it, the heavier liquid from higher in the outer core falls toward the inner core and pushes up the lighter alloy. The result is a **convection current** like those that occur in the oceans and in the atmosphere. Iron and its alloys are highly magnetic, so the moving currents of iron produce the magnetic field that extends from the outer core through the Earth and into the space around the planet.

convection current movement of a fluid (liquid or gas) caused when denser portions of the fluid move downward because of gravitational attraction, pushing less dense portions up and away

At the very center of the Earth, the density is the highest: 13.1 g/cm^3, or nearly 5 times as great as the density of the average crustal rocks. The

temperature at the center is estimated to be between 4,600 K and 6,600, but the pressure reaches 3.63 million atmospheres, keeping the inner core frozen solid. SEE ALSO ATMOSPHERE, EARTH'S; CONTINENTAL CRUST; EARTH MATERIALS, CHEMISTRY OF; GEOSPHERES; HYDROSPHERE; LITHOSPHERE; MANTLE; METEORITES; PLANETOLOGY, COMPARATIVE.

Earth, Views Of

In the 1960s and 1970s, early astronauts brought back photographs of Earth, and people first saw the planet as it looks hanging in the darkness of space. Like many new scientific observations, these images changed people's ideas: Earth, thought of for centuries as gigantic and indestructible, suddenly seemed small and fragile, its great "ocean" of atmosphere only a thin blue line.

Those photographs were examples of remote sensing—the science of collecting information by observing from a great distance. Remote sensing devices have been observing Earth from space since around 1960, when the first weather satellite was launched. They have changed the way people look at Earth, and they have helped spawn a new science.

Remote Sensing Devices.   Remote sensing systems are either passive or active; each type has its own best uses. Passive systems simply observe Earth and make a record of what they "see." Cameras are sensitive to visible light; they produce photographs that show light, dark, and natural color. They have been used, for example, to observe rock formations on Earth and the Moon. Infrared systems are sensitive to heat, so they produce images that show varying temperatures. They can map ocean currents such as the Gulf Stream, which is warmer than the ocean around it. Microwave sensors detect natural microwave emissions. They can tell observers, for example, where ice and snow lie, and they can help predict global warming through observations at the poles.

The Apollo 11 Lunar Module flies above the Moon's surface, seen from the Command/ Service Module. A half-Earth rises beyond the lunar horizon (July 21, 1969).

Ozone over Antarctica

Ozone (a special form of oxygen) in Earth's upper atmosphere filters out ultraviolet (UV) radiation—a dangerous form of radiation from the Sun. Remote sensors have tracked levels of atmospheric ozone since 1978. They indicate that the amount of ozone is decreasing, particularly over the polar regions. Increased UV radiation reaching Earth may threaten human health—UV rays can cause skin cancer; agriculture—UV rays can damage certain crops; and the ocean's supply of plankton—UV rays can penetrate the upper layers of ocean waters, damaging phytoplankton.

oceanography   science that deals with oceans

global warming   theory that suggests a gradual warming of the Earth's atmosphere as a result of increasing air pollution

greenhouse gas   gas in the atmosphere that traps heat and reflects it back to Earth

meteorologist   scientist who studies the atmosphere, especially with regard to climate and weather

seismologist   scientist who studies earthquakes

plate tectonics   theory used to explain continental drift (the movement of continents over the surface of the Earth) and other geological processes

earth systems science   field that integrates earth sciences such as geology, meteorology (science that deals with the atmosphere), and seismology (science that deals with earthquakes) to study the Earth as a whole

Active systems, such as radar and other microwave systems, send out their own pulses of energy and then measure smaller pulses that bounce back, much as echoes bounce back. The measurements are fed into a computer as a series of numbers, and the computer translates these into images or other forms that people can use. Radar, whose name stands for "radio detecting and ranging," can "see" through clouds, darkness, and water. It has been used to describe the planetary surface beneath the thick clouds of Venus and to map the Earth's ocean floors.

History.   People have always understood the value of a view from above—whether from the top of a hill (to scan for game) or from the turret of a fort (to sight an enemy). Remote observation was among the earliest uses of hot air balloons and airplanes. Photographs were taken from balloons as early as 1858 (in Paris). Aerial photography from planes started in 1909—just 6 years after the first airplane flew under its own power. More than 30 years later, World War II bombers carried radar systems that were forerunners of the modern radar-imaging systems flown on aircraft, satellites, and spacecraft.

Applications in Earth Science.   Observation devices in orbit have produced millions of new views of Earth. Many of these devices have everyday applications—for instance, those that track brush fires in Montana or the paths of hurricanes in the Atlantic and Pacific Oceans. Other remote sensing technology supports research in every branch of earth science and in many related sciences.

In agriculture and forestry, infrared satellite images can monitor the extent, type, and health of vegetation cover. Such images have been used to predict crop yields and to estimate springtime runoff of water from melting snow—water that will be available later for irrigation. Scientists in the field of **oceanography** use remote observation to estimate amounts of plankton—minute plant and animal life that floats or swims at or near the surface of the seas—in various parts of the ocean. These data help predict everything from commercial fishery catches (plankton is the basic foodstuff of the ocean) to **global warming** (phytoplankton—tiny plants—absorbs carbon dioxide, a **greenhouse gas**).

In geology, satellites have been used to study volcanoes and earthquake patterns, to search for minerals in the Earth's crust, and to prospect for oil. Satellite images like those on television weather reports help **meteorologists** study the causes and patterns of weather systems and related natural disasters such as floods. Satellites even help pinpoint fault lines and fracture zones—places where earthquakes are most likely to occur. Interpreted by **seismologists,** the results help engineers decide where to place roads, pipelines, and urban developments, and they add to scientists' understanding of movements in the Earth's crust (**plate tectonics**).

Earth Systems Science.   Views of Earth from space have made it possible to observe and monitor distant parts of the world's oceans, deserts, and polar regions; plant life in water and on land; the chemistry of the atmosphere; and even the changing elevation of the oceans. Using this "big picture," scientists are building computer models of the Earth's behavior—but to do so, they need to cross traditional scientific boundaries. For instance, in earth science, it is no longer enough to understand water, rock, or air systems by themselves, because the three interact. That is why remote views of Earth are contributing to a more inclusive field called **earth systems science.**

Earth systems science helps bring the separate disciplines together. One of its key uses in the future will be to study the global environment. For many years, remote sensors have tracked local environmental issues—pollution in Japan's Osaka Bay, whale populations in the Atlantic, or oil spills in Delaware. Earth systems scientists will look at worldwide issues, such as the thinning **ozone layer** on Earth's upper atmosphere, global warming, and the effects of ocean temperature changes—such as those caused by the periodic climate cycle called El Niño—on worldwide weather patterns. SEE ALSO CLIMATE CHANGES; GEOSPHERES; OZONE DEPLETION; PLATE TECTONICS.

Earth as a Dynamic System

ozone layer layer of the atmosphere that contains a special form of oxygen known as ozone

biosphere all of Earth's regions that support life

cryosphere frozen water in all its forms on the surface of the Earth

magnetosphere region above and around Earth that is affected by Earth's magnetism

phenomenon (*pl.* phenomena) fact or event that can be observed

greenhouse gas gas in the atmosphere that traps heat and reflects it back to Earth

biomass living and nonliving plant matter

Scientists who study Earth as a dynamic system observe how the six component parts of that system interact with one another. Those six parts are the solid Earth, the oceans, the atmosphere, the **biosphere**, the **cryosphere**, and the near space environment, or **magnetosphere**. Each component part is a system in itself, but the linkages among the systems give rise to a number of physical, chemical, and biological **phenomena** that affect the whole planet and its development. Given recent observations of global climate changes and other environmental concerns, it has become increasingly important to understand as much as possible about the planet's form and function—that is, Earth as a dynamic system.

Global Warming. An excellent illustration of a phenomenon resulting from the interaction among Earth's component parts is the process of global warming. Examination of surface temperatures over the past hundred years shows that the mean surface temperature worldwide has increased by about 0.5°C. Although scientists disagree as to the cause of the temperature increase—for example, whether it is a result of natural forces or of human activities—few seriously doubt that it has occurred.

Experiments with models of the atmosphere indicate that increased concentrations of **greenhouse gases,** mainly carbon dioxide (CO_2) and methane, can raise air temperatures. The industrial revolution began little more than 200 years ago. Since that time, factories, cars, trains, and other mechanisms that burn fossil fuels have been releasing increasing quantities of greenhouse gases into the air. For the past several decades, a number of weather stations scattered across the globe have carefully tracked the increasing concentrations of CO_2 in the atmosphere. When scientists include the increasing levels of greenhouse gases in their atmospheric models, they are able to approximate the 0.5°C surface warming of the 1900s.

Before the models could verify the observed temperature increase, however, scientists had to factor in two moderating influences: the oceans and that sulfate particles suspended in the atmosphere. The warming of ocean waters draws off some of the heat of increasing surface air temperatures. Sulfate particles in the atmosphere (from industrial activities and **biomass** burning) scatter solar radiation back to space, reducing the amount of solar heat that reaches the lower atmosphere. Both the ocean effect and the atmospheric sulfates might produce atmospheric cooling. They interact, however, with the warming effect of the greenhouse gases, moderating the warming trend in surface air temperatures.

Rise in Sea Level. Given rising global temperatures, there are good reasons to expect that the mean sea level would also rise. This, in fact, appears to be happening. Records of tide measurements that reach back more than 50 years strongly suggest that sea level is rising at annual rates measurable in millimeters. Two factors could account for the change: (1) melting of polar ice

Lava flowing into the ocean from the Kilauea Volcano, Hawaii Volcanoes National Park, illustrates heat transfer from the upper mantle to the crust and atmosphere.

and continental glaciers that increases the amount of seawater; (2) heat expansion that increases the volume of existing seawater.

Scientists working with the tide measurement data must also take into account long-term adjustments in the solid Earth structure in response to major ice-age glacial advances and retreats. The most recent glacial retreat began about 20,000 years ago and ended about 6,000 years ago. At the peak of the previous glacial advance, global sea level was reduced by well over 100 m. Water from ocean basins was frozen into the vast ice sheets that covered Antarctica and parts of the northern hemisphere. The melting of the ice sheets 6,000 years ago still affects tide measurements. This is true because the Earth's shape continues to change as it rebounds from the massive shifts of weight over its surfaces caused by glacial melting.

If the tide measurement data are not adjusted to leave out the effects of Earth's changing shape on sea level changes, the data vary greatly from place to place. This variation makes it difficult to see a connection between rising sea level and rising global temperatures. When the tide data are filtered so as to disregard the effects of the planet's changing shape, the variations in the global tide data are reduced. This is an excellent example of how the solid Earth component may influence long-term processes.

Changes in the Carbon Cycle.

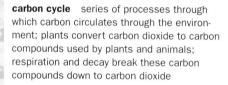

carbon cycle series of processes through which carbon circulates through the environment; plants convert carbon dioxide to carbon compounds used by plants and animals; respiration and decay break these carbon compounds down to carbon dioxide

Modern scientists have detected a further environmental mystery in the **carbon cycle.** As with the rise in sea level, the problem of the carbon cycle is linked with the issue of global warming. Researchers predict that the increasing atmospheric levels of CO_2 will have an impact on the climate. Even without human activity, the concentration of CO_2 in the atmosphere would be changing.

Atmospheric content levels of CO_2 have been falling for the last several tens of millions of years. It is thought that the atmospheric cooling that would naturally result from a reduction in CO_2 was a major cause of the last ice age. The same process may also account for the earlier ice ages that have marked the planet for the last million years in cycles of 100,000 years.

Scientists are most interested in identifying the one or more sinks—storage points—where CO_2 is being removed from the atmosphere. They must consider this factor in order to understand why the increased atmospheric concentration of CO_2 has been significantly less than would be expected from the known release of greenhouse gases by human activities, such as the burning of fossil fuels, wood, and other biomass. Models of the rate at which oceans absorb CO_2 show that the oceans alone could not account for the storage of carbon. It has been suggested that the continental biosphere must be expanding, thereby increasing the amount of CO_2 stored by land plants. Again, understanding a single phenomenon—changes in the carbon cycle—is linked to an understanding of the interactions of dynamic components: the atmosphere, the oceans, and the biosphere.

Ice Age Cycles.   As suggested above, the carbon cycle is involved in the ice age cycles that have dominated Earth history for the last million years. Core samples drilled from deep-sea levels of **sedimentary rock** and from deep ice in Antarctica have enabled scientists to determine the lengths and major events of these recurring cycles. In each cycle, the continental ice sheets have advanced for approximately 90,000 years and retreated for about 10,000 years, thus completing a cycle in about 100,000 years.

These findings are somewhat puzzling because the timing does not match the timing of variations in the solar radiation that reaches the Earth. Due to slight, predictable changes in Earth's orbit around the Sun, the amount of heat reaching Earth increases and decreases over cycles of 41,000 years. The ice age cycles of 100,000 years cannot be clearly related to the 41,000-year cycles in solar radiation.

Recent research has attempted to explain the time discrepancy. These studies indicate that three dynamic interconnections in the Earth system may be responding to variations in solar radiation in ways that result in the longer, 100,000-year cycles in the cryosphere. All three interconnections lengthen and/or shorten parts of the cycle. The first interconnection involves the carbon cycle. During glacial conditions, lower levels of CO_2 in the atmosphere contribute to climate cooling. Rising CO_2 levels during periods of glacial melting contribute to climate warming.

The second interconnection involves the deep, cold waters of the northern Atlantic Ocean. There is clear evidence of changes in the circulation pattern of these waters during glacial conditions that would further reduce atmospheric warming. Such a reduction encourages climate cooling, helping to extend the duration of the ice age. The third interconnection has to do with dust in the atmosphere. Periods of extremely high levels of atmospheric dust precede periods of glacial melting. If the dust load were to decrease the ability of ice sheets to reflect solar radiation, they would retain more warmth and melt more quickly. These feedback connections serve to extend the cycle and lengthen glaciation periods; then speed glacial melting once it begins.

Global Cooling and Its Effects.   The planet as a whole has been cooling since it took shape, approximately 4.55 billion years ago. Over the same general timescale, the solid Earth component has played a major role in the interacting planetary system.

The most important solid Earth issues are related to the processes that govern the rate of planetary cooling and the processes that result from the cooling. Essential to these processes is the ongoing heat transfer in the Earth's **mantle** because that transfer controls the rate of heat loss in the planet's interior. This interior circulation of heat is also connected with movements of the planet surface, such as **continental drift.** Modern studies of the mantle suggest that heat circulation occurs in two layers that are separated by a heat-absorbing layer located at a depth of 660 km. It is possible that intense flows of cooler materials

sedimentary rock   rock formed from deposits of sediments (soils, rock particles, and other materials) over long periods of time

mantle   region of the Earth between the molten core and the outer crust

continental drift   theory that the continents move over the surface of the Earth on large segments of Earth's crust and uppermost mantle (region just below the crust)

from the heat-absorbing layer into the lower mantle may periodically disrupt this heat layering.

The full role of the mantle's heat circulation in the cooling process remains to be seen. It is very clear, however, that the explanation of the cooling of the planet as a whole is linked to an understanding of the formation of the magnetic field shortly after the formation of the Earth. It is currently believed that the outward thrust produced by the growth of the Earth's solid inner core serves to trigger the process in the liquid outer core maintains the magnetic field. This triggering, however, would not come into play if the Earth as a whole were not cooling. Thus the fluid outer core, the overlying more solid mantle that controls the rate of planetary cooling, and the magnetosphere are interconnected. SEE ALSO CLIMATE CHANGES; GEOSPHERES; HEAT BUDGET OF THE EARTH; ISOSTASY; MAGNETIC FIELD, EARTH'S; MANTLE; PLATE TECTONICS.

Earth Materials, Chemistry of

The Earth was formed some 4.55 billion years ago from a cloud of gas and dust called the solar nebula, which resulted from the explosion of one or more stars in this region of the Milky Way galaxy. These explosions, called supernovas, scattered into space the chemical elements that eventually came together to form the planets in the solar system. All of the materials that make up the Earth—rocks, air, water, and all living things—are composed of those elements. By studying the chemistry of Earth materials and the chemical reactions and geological processes that create and destroy them, scientists can better understand the structure and workings of the Earth.

Chemical Makeup of the Solar System

Of all the objects in the solar system, the Sun is believed to be chemically most similar to the solar nebula from which the solar system formed. Scientists study the Sun to determine its chemical composition by examining the wavelengths of energy in sunlight. In addition to information from the Sun, chemical analysis of **meteorites,** comets, and Earth rocks can provide insight into the chemical makeup of the solar system. Some of the meteorites studied are rocks, found mainly in Antarctica, that have fallen to Earth after being knocked from the Moon and Mars by meteorite impacts. Chemical data are also gathered from rocks brought back from the Moon.

meteorite solid body that has fallen to Earth from space

These studies indicate that the most abundant elements in the solar system are hydrogen and helium (although these elements are not as plentiful on Earth, the Moon, and Mars, due to the explosive chemical changes that took place during the formation of the solar system). Studies also show that lighter elements are generally more abundant than heavier elements. This information helps scientists to understand the nuclear reactions that occurred in supernovas to produce the solar nebula. The relative abundance of the elements in the solar system also helps scientists recognize the effects of different chemical makeups on the planets in the solar system.

Creation of Atoms and Radioactivity

atom smallest, indivisible particle of an element

Nuclear reactions inside stars generate energy that is radiated into space, and produce about 2,500 different kinds of **atoms.** These atoms are distinguished

Periodic table of the elements.

PERIODIC TABLE

	I	II								
s	3 LITHIUM **Li** 6.941	4 BERYLLIUM **Be** 9.012								
	11 SODIUM **Na** 22.990	12 MAGNESIUM **Mg** 24.305								1 HYDROGEN **H** 1.008
d	19 POTASSIUM **K** 39.1	20 CALCIUM **Ca** 40.08		21 SCANDIUM **Sc** 44.966	22 TITANIUM **Ti** 47.88	23 VANADIUM **V** 50.942	24 CHROMIUM **Cr** 51.996	25 MANGANESE **Mn** 54.938	26 IRON **Fe** 55.847	
	37 RUBIDIUM **Rb** 85.47	38 STRONTIUM **Sr** 87.62		39 YTTRIUM **Y** 88.906	40 ZIRCONIUM **Zr** 91.224	41 NIOBIUM **Nb** 92.908	42 MOLYBDENUM **Mo** 95.94	43 TECHNETIUM **Tc** (98)	44 RUTHENIUM **Ru** 101.07	
	55 CESIUM **Cs** 132.90	56 BARIUM **Ba** 137.33		71 LUTETIUM **Lu** 174.967	72 HAFNIUM **Hf** 178.49	73 TANTALUM **Ta** 180.948	74 TUNGSTEN **W** 183.85	75 RHENIUM **Re** 186.207	76 OSMIUM **Os** 190.2	
	87 FRANCIUM **Fr** (223)	88 RADIUM **Ra** 226.025		103 LAWRENCIUM **Lr** (260)	104 RUTHERFORDIUM **Rf** (261)	105 DUBNIUM **Db** (262)	106 SEABORGIUM **Sg** (263)	107 BOHRIUM **Bh** (262)	108 HASSIUM **Hs** (265)	

f	57 LANTHANUM **La** 38.906	58 CERIUM **Ce** 140.115	59 PRAESEODYMIUM **Pr** 140.908	60 NEODYMIUM **Nd** 144.24	61 PROMETHIUM **Pm** (145)	62 SAMARIUM **Sm** 150.36	63 EUROPIUM **Eu** 151.965	64 GADOLINIUM **Gd** 157.25	65 TERBIUM **Tb** 158.925
	89 ACTINIUM **Ac** 227.03	90 THORIUM **Th** 232.038	91 PROTACTINIUM **Pa** 231.036	92 URANIUM **U** 238.029	93 NEPTUNIUM **Np** 237.048	94 PLUTONIUM **Pu** (244)	95 AMERICIUM **Am** (243)	96 CURIUM **Cm** (247)	97 BERKELIUM **Bk** (247)

proton positively charged particle in the nucleus, or center, of an atom

neutron uncharged particle in the nucleus, or center, of an atom

Global Warming

One serious problem identified by geochemists is the increased concentration of carbon dioxide in the atmosphere, caused by the burning of fossil fuels. Carbon dioxide causes the atmosphere to trap heat that would otherwise escape into space, leading to global warming that could change weather patterns. Long-term effects could include melting of the glaciers of Greenland and Antarctica, resulting in flooding of coastal areas. These coastal borderlands are home to approximately 70 percent of the world's population and account for much of its fertile farmland.

by the number of **protons** and **neutrons** contained in their nuclei, or centers. All atoms of a particular element have the same number of protons in their nuclei, but they may contain varied numbers of neutrons. Such variations are called isotopes of that element. For example, all carbon atoms have six protons, but not all have the same number of neutrons. Those with six neutrons are known as carbon-12 isotopes (because they have a total of six protons plus six neutrons); those with eight neutrons form a radioactive isotope called carbon-14. All the isotopes of an element have the same chemical properties, but those with more neutrons have a greater mass and are thus heavier.

Most of the atoms produced in stars are unstable, and over time their nuclei decay, releasing charged alpha (positive) and beta (negative) particles and high-energy gamma rays and producing more stable atoms. By the time Earth had formed, most of the unstable atoms had already decayed, so Earth is composed primarily of the stable atoms that make up the familiar elements. However, a few elements still have unstable, or radioactive, isotopes that decay so slowly that they have not yet been transformed into stable atoms. These elements include uranium, plutonium, radium, and a few others. Although radioactive atoms release gamma rays and nuclear particles that are harmful to life, they play an important part in Earth's chemistry. Radioactivity releases energy that is a vital source of heat within the Earth, and radioactive decay makes it possible to determine the age of various Earth materials, allowing scientists to assign dates to distinct geologic events and periods.

OF THE ELEMENTS

					VIII
					2 HELIUM **He** 4.003
III	IV	V	VI	VII	
5 BORON **B** 10.811	6 CARBON **C** 12.011	7 NITROGEN **N** 14.007	8 OXYGEN **O** 15.999	9 FLUORINE **F** 18.998	10 NEON **Ne** 20.180
13 ALUMINUM **Al** 26.982	14 SILICON **Si** 28.086	15 PHOSPHORUS **P** 30.974	16 SULFUR **S** 32.066	17 CHLORINE **Cl** 35.453	18 ARGON **Ar** 39.948

p appears to the left of the boron column.

27 COBALT **Co** 58.933	28 NICKEL **Ni** 58.69	29 COPPER **Cu** 63.546	30 ZINC **Zn** 65.38	31 GALLIUM **Ga** 69.723	32 GERMANIUM **Ge** 72.61	33 ARSENIC **As** 74.922	34 SELENIUM **Se** 78.96	35 BROMINE **Br** 79.904	36 KRYPTON **Kr** 83.80
45 RHODIUM **Rh** 102.906	46 PALLADIUM **Pd** 106.42	47 SILVER **Ag** 107.868	48 CADMIUM **Cd** 112.411	49 INDIUM **In** 114.82	50 TIN **Sn** 118.71	51 ANTIMONY **Sb** 121.75	52 TELLURIUM **Te** 127.60	53 IODINE **I** 126.905	54 XENON **Xe** 131.29
77 IRIDIUM **Ir** 192.22	78 PLATINUM **Pt** 195.08	79 GOLD **Au** 196.967	80 MERCURY **Hg** 200.59	81 THALLIUM **Tl** 204.383	82 LEAD **Pb** 207.2	83 BISMUTH **Bi** 208.980	84 POLONIUM **Po** (209)	85 ASTATINE **At** (210)	86 RADON **Rn** (222)
109 MEITNERIUM **Mt** (266)	110 UNUNILIUM **Uun** (269)	111 UNUNUNIUM **Uuu** (272)	112 (?)						

66 DYSPROSIUM **Dy** 162.50	67 HOLMIUM **Ho** 164.93	68 ERBIUM **Er** 167.26	69 THULIUM **Tm** 168.934	70 YTTERBIUM **Yb** 173.04
98 CALIFORNIUM **Cf** (251)	99 EINSTEINIUM **Es** (252)	100 FERMIUM **Fm** (257)	101 MENDELEVIUM **Md** (258)	102 NOBELIUM **No** (259)

Chemical Properties of the Elements

Elements can be classified in several ways, according to their chemical properties. The most common method of classification is based upon the number of protons in their nuclei.

The Periodic Table of Elements. In order to balance out the positive charge of the protons in its nucleus, an atom must also contain an equal number of negatively charged electrons. Since electrons participate in chemical bonding, they define the primary chemical properties of elements. To understand these chemical properties, it is important to understand the nature and action of electrons.

In 1913 the Danish physicist Niels Bohr first explained how the hydrogen atom works. His model showed electrons as small particles orbiting the nucleus of an atom, much as a planet orbits the Sun. However, physicists realized later that electrons did not behave exactly the way Bohr described, and the theory of **quantum mechanics** helped to explain electron behavior more precisely. Quantum mechanics theorizes that electrons actually distribute themselves among, and can move between, several available **orbitals.** The precisely ordered changes in the chemical properties of the various elements are a result of the way electrons distribute themselves among the orbitals.

Quantum mechanics has provided sufficient insight into the chemical properties of elements to permit construction of the periodic table. The table

quantum mechanics branch of physics that deals with the structure and behavior of atoms and subatomic particles

orbital specific area around the nucleus, or center, of an atom in which one or more electrons are found

ion atom or molecule that has a positive or negative electric charge

smelt to melt or fuse ores (metal-bearing minerals or rocks) and separate out the metals they contain

magma molten rock from deep within the Earth

mantle region of the Earth between the molten core and the outer crust

volatile easily vaporized at moderate temperatures and pressures

sediment soils, rock particles, and other materials that are deposited over time and make up the ground, whether on dry land or at the bottom of a body of water

groups together elements that have similar chemical properties. For example, one group of elements, called the alkali metals, includes lithium, sodium, potassium, rubidium, and cesium. When dissolved in water, all of these elements form positive **ions** with an electrical charge of +1. Similarly, the halogens (fluorine, chlorine, bromine, and iodine) form negative ions with an electrical charge of –1, and the noble gases (helium, neon, argon, krypton, xenon, and radon) do not form ions at all. When elements from the same group are acted upon by geological processes, they tend to remain together. For example, rocks that contain high concentrations of potassium commonly also contain other alkali metals, such as rubidium and cesium.

Geochemical Classification. Elements can also be classified according to the type of Earth materials in which they are most often concentrated. When ores are **smelted** to recover metals, three different liquids are produced: liquid iron, liquid sulfide, and liquid silicate. In addition, gases are released into the atmosphere when ores are smelted. Certain elements concentrate in specific liquids; others associate with the gases. For example, nickel, gold, and tin tend to concentrate in liquid iron, whereas copper, silver, and lead are found in liquid sulfide. Interestingly, oxygen tends to associate with liquid silicate, which means that it is concentrated in rocks rather than in the atmosphere.

The grouping of elements in this fashion is called geochemical classification and helps to clarify how Earth formed. Possibly, when Earth first formed, it was completely molten, and large masses of liquid iron sank to the center to form the core. Silicate liquid surrounded the core as a **magma** ocean that eventually formed the **mantle** and the crust. While the sulfide liquid also sank, it did not form a continuous layer between the mantle and the core. The elements that concentrate in gases (like hydrogen and nitrogen), together with **volatile** compounds like water and carbon dioxide, escaped the magma ocean and formed the atmosphere.

Applying Geochemical Knowledge

Understanding the Earth's chemistry can help scientists solve many practical problems. For example, geochemical classification helps explain why only nine elements (all elements that concentrate in liquid silicate) make up nearly 99 percent of the weight of Earth's crust: oxygen, silicon, aluminum, iron, calcium, magnesium, sodium, potassium, and titanium. All of the other elements combined make up about 1 percent, including such valuable industrial metals as copper, tin, zinc, lead, silver, and gold. This scarcity makes finding such valuable elements, planning how to remove them from the Earth, and disposing of the waste products resulting from their use a challenge to earth scientists.

To carry out these tasks, geochemists study processes such as the rock cycle and the hydrologic (water) cycle. In the rock cycle, magma from the mantle cools to form igneous rocks. These rocks are eroded over time to form sediments that are deposited and eventually compressed to form sedimentary rocks. High temperatures and pressure created during the formation of mountains convert sedimentary rocks into metamorphic rocks. Under extreme conditions, these rocks may melt to form new magma, beginning the cycle again.

In the hydrologic cycle, water from the surface of the oceans evaporates into the atmosphere, where it is transported to the continents in the form of rain, hail, or snow. Rainwater, in turn, transports **sediment** and dissolved ions; it also promotes chemical reactions among the minerals with which it

Geologic cycle by which chemical and physical processes transform magma to igneous rocks, then to sedimentary and metamorphic rocks, and ultimately back to magma.

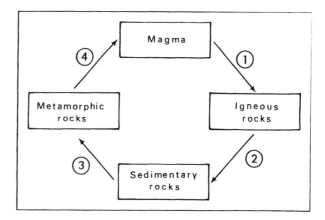

comes into contact. Thus the circulation of water is part of the other processes that affect rocks and make life possible on Earth.

Four areas of the Earth—the lithosphere (rocks), the hydrosphere (water), the atmosphere (air), and the biosphere (living things)—interact with one another to produce and support life. Scientists study these interactions to predict how they will respond to environmental challenges. The apparent impacts of some of these challenges, such as pollution from factories, may be local, but all such challenges have an overall effect on the world's environment. For this reason, geochemists now monitor the oceans, land, and atmosphere to detect contamination and other changes that may be harmful to life. The discovery of the deterioration of the **ozone layer** over the polar regions is one example of this important aspect of **geochemistry**. SEE ALSO EARTH, ORIGIN OF; EARTH AS A DYNAMIC SYSTEM; GALAXIES; GEOCHEMISTRY; GEOSPHERES; GOLDSCHMIDT, VICTOR MORITZ; HEAT BUDGET OF THE EARTH; IGNEOUS ROCKS; MAGMA; MANTLE; METAMORPHIC ROCKS; METEORITES; ORE; OZONE DEPLETION; SEDIMENTS AND SEDIMENTARY ROCKS; SOLAR SYSTEM; STARS.

ozone layer layer of the atmosphere that contains a special form of oxygen known as ozone

geochemistry science that deals with the chemical composition of the Earth's materials

Earth-Moon System, Earliest History

erosion wearing away of land by wind and water

plate tectonics theory used to explain continental drift (the movement of continents over the surface of the Earth) and other geological processes

meteorite solid body that has fallen to Earth from space

mantle region of the Earth between the molten core and the outer crust

volatile easily vaporized at moderate temperatures and pressures

Until quite recently, scientists had almost no understanding of the first third of Earth's history because geologic processes such as **erosion** and **plate tectonics** had destroyed the planet's oldest rocks. However, data gathered from moon rocks and **meteorite** studies have revealed much about events that led to the formation of the Earth some 4.55 billion years (Ga) ago.

The planets formed when cosmic dust grains lost energy as they collided, then aggregated (stuck together) because of gravitational attraction. Asteroid-sized bodies called planetesimals formed within 20 million years (Ma) of the formation of the Sun, and Earth probably reached its present size within 50 Ma. Current theories suggest that the Moon was created soon afterward, when a planetesimal the size of Mars struck the Earth, blasting material from the rocky **mantles** of both the Earth and the planetesimal into orbit around Earth. Continuing bombardment by planetesimals played an important role in the early history of the Earth and Moon, probably melting a thick outer layer of both bodies.

The early atmosphere of Earth was most likely formed when, as bombardment heated the surface, iron separated and moved inward to form a core, and **volatile** materials such as water and carbon dioxide (CO_2) were released, creating the atmosphere. As the Earth cooled, water condensed to form the oceans, leaving a thick CO_2 atmosphere. Much of the CO_2 eventually dissolved into the oceans, creating a weak acid solution that attacked deep-sea rocks, causing much of the CO_2 to become "stored" in the rocks. When plants came into being, about 2.5 Ga ago, they consumed more of the CO_2 and released oxygen, creating the oxygen-rich atmosphere that led to the development of other life-forms.

Exactly when the first life on Earth arose is a matter of debate among scientists. Some believe that it occurred soon after Earth's creation; others say that early impacts delayed its formation for several hundred million years; still others argue that it may have started, been extinguished by impacts, and restarted several times. This last view has been supported by evidence of major **extinctions** caused by large-scale impacts as recently as 66 Ma. SEE ALSO EARTH, ORIGIN OF; EXTINCTIONS; GRAVITY; MOON; PLATE TECTONICS; WEATHERING AND EROSION.

extinction dying out completely, leaving no direct descendants (refers to a specific grouping of organisms)

Earthquakes

An earthquake, one of the most dreaded natural events, is a shaking of the ground caused by seismic waves—vibrations passing through Earth. Seismic waves could also be called shock waves. The prefix *seis-*, in fact, comes from the Greek word *seismos,* meaning "shock," and appears in other terms related to earthquakes. Seismology, for example, is the scientific study of earthquakes. Seismicity is the frequency, strength, and geographic distribution of earthquakes.

Causes of Earthquakes

Seismic waves may have a variety of natural or human causes, such as the movement of liquid rock in volcanoes; nuclear or chemical explosions; and rock breakage in mines. By far the most common (and most energetic) source of seismic waves, however, is **plate tectonics.**

plate tectonics theory used to explain continental drift (the movement of continents over the surface of the Earth) and other geological processes

Movement Along Faults. The theory of plate tectonics supposes that the upper 75 km or so of the solid Earth, the lithosphere, is divided into plates—large, rigid units that move with respect to each other. The movement may involve plates rubbing against one another, pulling apart, or sliding under and over each other along their edges. For example, the Pacific plate rubs against the North American plate in the area of the San Andreas **Fault** in California.

Seismologists believe that plate movements produce stresses in the solid rock layers of the lithosphere. The rocks absorb this tectonic stress until it becomes great enough to make the rocks strain, bend, and suddenly break and slide—in much the same way that a ruler bends and finally snaps when enough pressure is applied. As the rocks snap back after rupture, they send out seismic waves, which cause earthquakes.

fault break or crack in the Earth's rock layers that creates a surface along which the rocks slide

seismologist scientist who studies earthquakes

Rock layers break and slide along faults. The direction of the rock slippage—horizontal, slanted, or vertical—depends on the direction of plate movement. The extent of the slippage may vary, amounting to as much as several meters instantaneously in the case of great earthquakes. The point at which rock movement sends out seismic waves is known as the hypocenter, or focus, of the earthquake. The point on the Earth's surface directly above the hypocenter is the epicenter. Most hypocenters are located within the first 75 km beneath the surface of the Earth. Some hypocenters, however, are located at points where tectonic plates are sliding over and under each other at depths as great as 700 km.

Types of Seismic Waves. There are two main types of seismic waves: surface waves and body waves. Surface waves occur near the surface of the Earth.

Damage to the Oakland–San Francisco Bay Bridge from the October 1989 earthquake.

Body waves travel outward from the hypocenter through the entire body of the Earth. Surface waves cause the greatest damage during an earthquake.

Seismic waves associated with large earthquakes travel through and completely around the Earth. The study of these seismic waves is the most powerful scientific tool for understanding the Earth's interior structure. Studies of this kind help to define the major divisions of the crust, mantle, and core and show that a liquid outer core surrounds Earth's solid inner core.

Size and Frequency

The first attempts to classify earthquakes focused on the degree to which an earthquake was felt. Seismologists later developed more scientific methods.

Modified Mercalli Scale.   Giuseppe Mercalli of Italy developed an early scale for measuring earthquakes. A modification of his system is still used in the United States. The modified Mercalli scale classifies the size of earthquakes on a scale of 1 to 12, usually given in Roman numerals (I–XII), on the basis of what people report they felt or heard, how much damage was done, and other effects. The Mercalli scale is useful for comparing the force of one earthquake to that of another, but it is not the most reliable method for measuring size, nor is it useful for measuring earthquakes that occur in unpopulated areas.

Richter Scale.   The Richter scale, developed in the 1930s by Professor Charles Richter at the California Institute of Technology, is more scientific and reliable. Using **seismographs** to measure the **amplitude** of seismic waves, Richter assigned a size value of 3 to an earthquake that would produce a maximum amplitude of 1 mm on a seismograph located 100 km from the epicenter. He then defined each tenfold increase or decrease in maximum amplitude as one whole step on his scale. Because seismic wave amplitudes decrease as

seismograph   machine for measuring and recording Earth movements such as tremors and earthquakes

amplitude   distance between the midpoint and the maximum height or depth of a wave

125

distance from the epicenter increases, the Richter scale can also measure the size of earthquakes at distances other than 100 km from the epicenter.

The Richter scale can accurately measure the size of earthquakes up to values of about 7 or 8. All great earthquakes, however, register near 8.5, although some release more energy than others. Newer seismographs are capable of more detailed and complete measurements of seismic waves, making it possible to describe the size of the most intense earthquakes more precisely.

Frequency of Earthquakes.   The minimum size on the Richter scale for earthquakes that can be felt is generally about 2.5 to 3.0. An estimated 1,000 earthquakes of this size occur every day, but most go unnoticed because they occur far from places where people live. The frequency of earthquakes diminishes with increasing size.

Locations of Earthquakes

Past experiences with earthquakes have long suggested that earthquakes occur much more commonly in some places than in others. By the 1960s, seismologists had realized just how well defined the earthquake belts of the Earth are.

Along Plate Boundaries.   Locating and charting Earth's major earthquake zones has been the work of the World Wide Standardized Seismograph Network (WWSSN), established in the late 1950s. At that time, the United States intended to use the network to detect nuclear weapons tests in foreign countries, but the network produced the side benefit of improving seismologists' ability to detect and locate earthquakes. Data from the WWSSN indicated that most earthquakes occur where **tectonic plates** meet and move in relation to each other. The pattern of plate boundaries corresponds with earthquake belts, or zones, winding over the surface of the planet. Some of the most active earthquake zones occur in China, Iran, the Mediterranean region, and the western edges of North and South America.

Midplate Earthquakes.   Although less frequent, earthquakes can and do occur in the midsections of tectonic plates. The faults that cause the seismic activity are rarely exposed on the surface, and their depth is seldom identified.

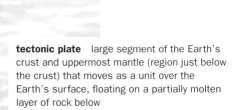

tectonic plate   large segment of the Earth's crust and uppermost mantle (region just below the crust) that moves as a unit over the Earth's surface, floating on a partially molten layer of rock below

Frequency of Earthquakes Based on Observations Since 1900

Description	Size	Yearly Average
Great	8 and higher	1
Major	7–7.9	18
Strong	6–6.9	120
Moderate	5–5.9	800
Light	4–4.9	6,200 (estimated)
Minor	3–3.9	49,000 (estimated)
Very Minor (less than 3)	2–3 1–2	about 1,000 per day (estimated) about 8,000 per day (estimated)

From National Earthquake Information Center QED tables.

Seismologists do not clearly understand the cause of midplate earthquakes, but they believe that these earthquakes do seem to occur in response to the same forces that cause plates to move.

Midplate earthquakes are often deadly because they hit in areas that are poorly prepared for such events. The 1993 earthquake that struck Latur, India, hit in an area with no known history of earthquakes. More than 11,000 people were killed. Three of the most powerful earthquakes in U.S. history, all greater than 8 on the Richter scale, occurred in a midplate zone along the Mississippi Valley. These earthquakes hit in 1811 and 1812 and were felt over nearly the entire eastern half of the nation. One of them caused church bells to ring in Boston, 1,000 miles from the epicenter.

Effects of Earthquakes

No natural disaster has a greater capacity than earthquakes for causing sudden death and destruction. The severity of earthquake disasters depends on a number of factors, some natural and some human.

Factors That Govern Damage. The size of an earthquake is the most obvious factor in severity of damage. The larger the earthquake, the more energy is released, resulting in larger ground motion and longer periods of a high degree of shaking.

Distance from the epicenter is the next most obvious factor. The amount of ground motion falls off rapidly as distance increases. At similar given distances in earthquakes of similar sizes, depth also becomes a factor. In general, the deeper an earthquake, the less its effects will be felt on the surface.

Often the extent of earthquake damage depends on the type of ground on which buildings are constructed. As a general rule, damage is greater in cities or settlements built over loose sediment or thick soil than in communities built over solid bedrock. Buildings on these less stable surfaces may collapse from

Damage in Tokyo caused by the September 1923 earthquake.

Earthquakes

Danger Underfoot in California

The San Andreas Fault runs in a roughly north-south direction across most of California and all of Baja California. This fault marks the active boundary between the Pacific plate and the North American plate. Horizontal movement of the Pacific plate in a northwestern direction builds up stress in the rocks along the fault. At least 19 strong or major earthquakes occurred in California in the 1900s, the worst in San Francisco in 1906. Some estimates place human casualties at 500 or more. A second, less intense shock struck San Francisco in 1989. Some scientists believe that the 1989 shock was a delayed aftereffect of the 1906 earthquake. Had the 1989 earthquake been as severe as the one in 1906, the death toll could have risen to 11,000.

the effects of the seismic waves themselves or from the liquefaction process, in which the water-soaked soil suddenly loses its ability to support the weight of buildings.

Other factors that may cause death, injury, and destruction include fire, which often follows an earthquake. Fire burned for three days in a 4-mile-square area after the 1906 San Francisco earthquake. An earthquake on the ocean floor may set off a tsunami, or tidal wave, that can swamp shorelines.

The Historical Record.   Perhaps the worst natural disaster known, not counting plagues and famines, occurred with the earthquake that struck central China in 1556. Most of the people in the region lived in caves carved from soft rock. These dwellings collapsed during the earthquake, killing an estimated 830,000 people. The official death toll of the 1978 earthquake in Tangshan, China, was 243,000, but many experts believe that the actual number may have been three times as high. The three other earthquakes with the highest tolls in human life occurred in Ningxia Province, China, in 1920 (200,000 dead); Aleppo, Syria, in 1138 (100,000 dead); and Tokyo, Japan, in 1923 (99,300 dead).

The United States has been rather fortunate with respect to earthquake disasters. Only the 1906 earthquake in San Francisco killed more than about 100 people.

Reducing Earthquake Hazard

Earthquakes can cause tremendous damage to people and property. Many efforts have been made to identify where earthquakes are likely to occur and find ways to prepare for them.

Locating Source Zones.   To reduce the hazard earthquakes present, it is necessary to first identify and describe the source zones of seismic activity. These zones may be known faults or simply areas where earthquakes have occurred (not all faults are historically active). Networks of seismograph stations in various regions have provided much knowledge of active source zones. The University of California at Berkeley set up the first regional network, consisting of 15 stations connected by telephone wires. Later, the U.S. Geological Survey (USGS) set up a system of 200 stations south of San Francisco. Many additional networks have since been established in California and elsewhere.

Building Design and Construction.   The surest way to prevent earthquake damage is to avoid constructing buildings in active earthquake zones. Where construction is unavoidable, it is possible to design and build structures that resist earthquake damage. The basic design principle is resistance to horizontal as well as vertical stress. Engineers continue to make much progress in this area of structural design.

Response Readiness.   Preparing for earthquakes that are bound to strike is another important aspect of damage reduction. This preparation requires that communities develop emergency response plans and provide special training for fire, police, and medical personnel. The general public must be informed about what to expect in an earthquake and what precautions to take in homes, schools, and workplaces.

Attempts to Predict Earthquakes

The topic of earthquake prediction probably excites as much general interest as the earthquakes themselves. Among scientists, the subject remains a challenge

Possible Indications. Seismologists and others have suggested that a number of observable events occur before an earthquake and so can be used as predictors. The list includes changes in the speed of body waves in small earthquakes; tilting of the ground surface; unusual electric or magnetic signals; changes in the flow of springs, the amount of radon gas in **groundwater,** and the level of water in wells; increase in the occurrence of small earthquakes; and unusual behavior of animals. People who have attempted to predict earthquakes in the past often used a combination of these indicators.

groundwater underground water that supplies natural springs and can be tapped by wells

Reliability of Current Efforts. Unfortunately none of the known methods of prediction has proved reliable. In 1975 some Chinese seismologists used four of the indicators listed above and successfully predicted an earthquake in the Manchurian province. Millions of people were evacuated, and only 300 died. A year later, Chinese seismologists—relying on similar indicators—failed to predict the Tangshan earthquake, the deadliest earthquake of the 1900s.

Seismologists continue to pursue the goal of accurate earthquake prediction. They hope that ongoing research into the processes surrounding faults will help unravel the complex workings of Earth and lead to the discovery of reliable warning signals of earthquakes. SEE ALSO EARTH, STRUCTURE AND COMPOSITION OF; EARTH'S CRUST, HISTORY OF; PLATE TECTONICS; RICHTER, CHARLES F.

Earth's Atmosphere

See Atmosphere, Earth's.

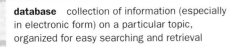

Earth Science Information

Researchers, scientists, teachers, students, and others can turn to a wide variety of materials for information about the earth sciences. Although much information is still stored in traditional printed form, computer technology has provided many new ways to store, obtain, and share information. Those who use earth science information must have the skills to use all kinds of formats, from journals to microfilm to on-line articles and File Transfer Protocol (FTP) sites. Some earth science projects create huge amounts of data, and managing these data is a key area of activity in the earth sciences. **Archivists** face a number of challenges in preserving accurate information for the future.

archivist person responsible for preserving and caring for stored data

Finding Information. A good place to start the search for earth science information is with indexes and catalogs in a library. Some catalogs may be in card or print format, but on-line catalogs have become standard because they are generally easier to use. The catalogs of many libraries are often linked in an electronic **database,** so that if a particular resource is not available at one library, it can be borrowed from another. Researchers with access to the Internet can browse the catalogs of libraries around the world from their own homes or offices or from library computer terminals. Librarians and information specialists are trained to help people use all types of catalogs to identify and locate primary information sources.

database collection of information (especially in electronic form) on a particular topic, organized for easy searching and retrieval

Someone who is researching a topic in the earth sciences may choose to begin with a general reference book such as a dictionary, encyclopedia, handbook,

Earth Science Information

or directory. There are many such reference books devoted specifically to the earth sciences, such as dictionaries of geology, handbooks of mineralogy, and the American Geological Society's *Directory of Geoscience Departments.* The supplementary readings listed at the end of this encyclopedia offer a good starting point.

Another search approach begins at a World Wide Web site. Many Internet browsers offer indexes of Web sites organized by topic. For example, the Yahoo browser at http://www.yahoo.com/Science/ offers indexes for Earth Sciences, Ecology, Geography, and Geology and Geophysics. Users of America Online can go to the AOL Science channel for an index to Environment and Nature as well as an index to Geology. Such sites are starting points from which the researcher can jump to thousands of Web pages devoted to a multitude of earth science topics. Many of them are links to the home pages of agencies that produce and store data, including **NASA, NOAA,** and the United States Geological Survey (USGS).

Electronic Data Storage. Increasing amounts of earth science information—including written text, data in numerical form, and images—are stored in electronic formats such as magnetic tapes for mainframe computers, CD-ROMs or floppy disks for personal computers, and the Internet. Many organizations exist simply to collect and manage earth science data. Examples include NASA's Planetary Data System and the Southern California Earthquake Center's SCEC Data Archive. The Earth Observing System Data and Information System (EOS-DIS) is one of many databases operated by the federal government. It offers access to hundreds of earth science information products and services.

Electronic storage allows users to manipulate data in ways that would be time-consuming and difficult with print media. For example, after a 1994 earthquake, the California Office of Emergency Services produced several maps combining language data from the most recent census, a street grid from a private mapping company, and damage information from local government agencies. The maps showed the buildings that had been damaged in the earthquake and the languages spoken by their inhabitants. The Federal Emergency Management Agency used the maps to assign field workers with the needed language abilities. Similarly, separate data sets for geology, fault locations, and **seismological** readings can be combined into a single map for use in assessing earthquake risk.

As more and more information is stored and communicated electronically in digital form, however, concerns arise about the accuracy of the data. Without a master copy or original to check, there is no definite way to be certain that numbers copied from one application and pasted into another are correct. Mapping and geographic information are of special concern because shifts in scale—from a map of a large area to a close-up, more detailed map of a smaller area, for example—can distort data.

Preserving Data. Unlike information in some other scientific fields, earth science information tends to remain useful even when it is hundreds of years old. Preserving such materials is one of the many challenges faced by modern archivists.

Old paper documents, such as books and journals, become brittle and start to fall apart. For years librarians have solved this problem by copying the documents onto microfilm for users to scan through special readers. However, microfilm technology cannot be used for oversized maps or for the color images that are so important in the earth sciences. Experts are investigating possibilities for new kinds of long-lasting color film and ways to handle large-sized

NASA National Aeronautics and Space Administration, the U.S. space agency

NOAA National Oceanic and Atmospheric Administration, a U.S. agency that coordinates earth science research

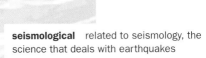

seismological related to seismology, the science that deals with earthquakes

130

materials. Even if such preservation technology becomes available, however, it may prove very expensive. Meanwhile, old paper resources are in danger of falling apart before they can be preserved in some permanent and usable form.

Electronic storage seems like an attractive method for preserving data and documents, but it too presents problems. Archivists need to keep material in a form that can be used for a long time, but the life spans of formats such as CD-ROM are not yet known. In addition, technology changes so fast that an electronic storage method can quickly become obsolete. Some valuable data sets stored on magnetic tape have already been lost because either the equipment to access them is no longer available or the software used to record them no longer exists.

Earth Sciences, Early History

natural history systematic study of natural objects, especially in their natural settings (includes physical and life sciences)

Although modern geology emerged as a field of study in the 1700s, the earth sciences have roots much farther back in history. From ancient times through the scientific revolution of the late 1500s and the 1600s, many thinkers and experimenters studied subjects such as astronomy, geography, and **natural history**—although it would be years before the complex links among these subjects became clear.

Early Breakthroughs. Many ancient civilizations developed high degrees of skill in mathematics and astronomy. The Babylonians and Chinese observed the sky closely and created maps of the heavens that allowed them to predict when eclipses would occur. Between 1200 and 100 B.C., the Olmec Indians of Mexico developed a calendar based on their astronomical observations, and centuries later the Maya Indians built observatories to study the motions of the stars and planets. The ancient Egyptians used astronomy to keep track of the seasons and also developed a system for land surveying.

The ancient Greeks were among the first to focus on separating scientific ideas from superstition. They not only observed features of the natural world but also tried to develop theories that would explain their observations. Aristotle (384–322 B.C.) was a giant among Greek thinkers. He studied and wrote about nearly every aspect of science, although biology was his chief interest. Aristotle developed a system of classifying animals and plants into organized categories and frameworks and came up with the idea that similar **species** might be related, although he never developed a theory to explain such relationships.

species narrowest classification, or grouping, of organisms according to their characteristics; members of a species can reproduce only with others of that group

optics science that deals with the properties of light; can include the use of mirrors and magnifying lenses

Ptolemy (ca. A.D. 100–ca. 170) was one of the foremost earth scientists of the ancient world. Although he wrote about many subjects, including **optics** and geography, he was best known as an astronomer. His most influential work, the *Almagest*, summarized the ideas of many Greek astronomers. Although Ptolemy correctly described the world as spherical, he incorrectly claimed that the Earth is at the center of the universe and that the Sun, stars, and planets revolve around it. People continued to believe in an Earth-centered universe for hundreds of years after Ptolemy's death.

The ancient Romans, who conquered Greece and much of the surrounding territory, did not engage in much scientific investigation. However, they did collect the scientific knowledge of their day in several large encyclopedias. Pliny the Elder (ca. A.D. 23–79) produced 37 volumes on the subject of natural history, and the historian Strabo (63 B.C.–ca. A.D. 25) created a 17-volume geography of the known world.

The Scientific Revolution. Early in the Middle Ages, from the A.D. 400s and for hundreds of years thereafter, science was of little concern to European scholars. In the 1000s, however, they began taking a renewed interest in science. Europeans founded their first universities in the 1100s. In centuries to come, these institutions would promote the growth of scientific knowledge.

The 1500s saw the beginning of a revolution in scientific thinking. It started in 1543, when a Polish astronomer named Nicolaus Copernicus (1473–1543) published a new theory of the universe in which he claimed that the Earth and other planets revolved around the Sun. The Copernican theory, which seemed to threaten traditional religious notions, deeply disturbed many people. Copernicus died shortly after his book was published, but other scientists were quick to pick up on his ideas.

The career of Italian scientist Galileo Galilei (1564–1642) was a landmark in the development of modern science. Galileo developed a high-quality telescope and made systematic studies of the heavens, discovering that there were many stars invisible to the naked eye and moons orbited the planet Jupiter. He was also a pioneer in devising experiments to test theories about the laws of physics. In 1632 Galileo offended the Catholic Church by publishing a book in which he defended the Sun-centered, Copernican view of the solar system. He was forced in 1633 under threat of execution to deny publicly any ideas that did not agree with the old Ptolemaic system.

German astronomer Johannes Kepler (1571–1630) was active in the scientific world at the same time as Galileo. Kepler put the final touch on the Copernican theory by demonstrating that the movements of the planets could be accurately accounted for and predicted if the planets moved around the Sun in elliptical (oval) orbits rather than circular ones.

English mathematician and scientist Isaac Newton (1642–1727) developed a theory that explained why all objects in the universe, from apples to moons, move the way they do. His 1687 book, *Principia Mathematica*, set forth the concept of universal gravitation, an invisible force that all objects with **mass** exert on other objects. Building on the work of Galileo, Newton formulated three laws of motion that became the basis for a new wave of studies of the Earth and solar system and for a new understanding of physics. SEE ALSO ASTRONOMY, OBSERVATIONAL; COSMOLOGY; EARTH, VIEWS OF; EARTH SCIENCE INFORMATION; GRAVITY.

mass amount of matter that causes an object to have weight when it is in an area influenced by gravity; commonly measured in kilograms

Earth Sciences, Research in

Research in the earth sciences can be either basic or applied. Basic research is aimed at answering theoretical questions, whereas applied research attempts to solve practical problems. However, the distinction between basic and applied research is not clear-cut. The same research that helps scientists understand the processes and forces that operate on and within the Earth often helps society, for example, to improve living standards and protect the environment.

Earth scientists investigate ways of protecting people from natural hazards such as floods, earthquakes, volcanoes, landslides, wildfires, and violent storms. They evaluate techniques for finding, conserving, and using resources such as energy sources, minerals, and water. Earth scientists also study the geologic past to help them predict future trends—for example, how the Earth will respond to climate changes.

Disciplines in the Earth Sciences. The earth sciences, also called the geosciences, link together many fields of study that often overlap with one another and with other branches of science and engineering. At the center of this interlinked network is geology, narrowly defined as the study of rocks,

Who Pays for Earth Science Research

Funding for research in the earth sciences comes from a variety of sources, including government, industry, private foundations, and universities. Many geoscientists work on practical rather than theoretical topics—applying geology and other earth sciences to issues involving natural hazards (such as floods or earthquakes), resources, and the environment. More often than ever before, earth scientists work on problem-solving teams with scientists in other fields as well as with engineers, social scientists, and the people who make laws and public policy.

tectonic plate large segment of the Earth's crust and uppermost mantle (region just below the crust) that moves as a unit over the Earth's surface, floating on a partially molten layer of rock below

faulting movements of the Earth's crust that produce a fault (a break or crack in the Earth's rock layers that creates a surface along which the rocks slide)

folding movements of the Earth's crust that cause rock layers to bend into folds

fault break or crack in the Earth's rock layers that creates a surface along which the rocks slide

sediment soils, rock particles, and other materials that are deposited over time and make up the ground, whether on dry land or at the bottom of a body of water

bedrock solid rock beneath soil and other surface cover

but often interpreted to include any study of Earth—in short, all of the geosciences.

Any application of basic science—physics, chemistry, or biology—to the study of Earth can be considered part of the earth sciences. For example, geophysics (the study of Earth's physical characteristics and structure) and meteorology (the study of Earth's atmosphere, climate, and weather) are applications of physics. Geochemistry (the study of the chemical makeup of the Earth's materials), mineralogy (the study of minerals and their properties), and metallurgy (the study of metals) are applications of chemistry. Paleontology (the study of prehistoric life) is related to biology. There are many other scientific fields, some broad and some highly specific, that contribute to the understanding of the highly complex problems that geoscientists face.

Scope of the Earth Sciences. An earth scientist's laboratory is the whole planet—and other planets and moons (and even other solar systems), whenever they can be closely observed through telescopes or examined through actual samples of their substance. Geoscientific investigations range in scale from microscopically small to enormous. One earth scientist may specialize in producing highly magnified images of the atoms on the surfaces of minerals; another may search for signs of life throughout the entire known universe.

The geosciences have a wide scope in time as well as in size. Many geological processes, such as the formation of mountains, the movement of **tectonic plates,** and the sculpting of the landscape, take place over thousands or even many millions of years. Rocks on the Earth and on the Moon record events that took place nearly as far back in time as the origin of the planet itself, about 4.55 billion years ago. Studying processes that occurred on this kind of expanded timescale is called working in deep time.

Geoscientists have developed several methods for studying deep time. Some techniques are used in the field, such as geologic mapping, which helps unravel questions about the timing of layering, tilting, **faulting,** and **folding** of rocks. Other methods involve laboratory work—for example, measuring the amount of radioactive material in a rock to determine its age.

Not all geological processes happen on such vast timescales, however. Many earth scientists study events, such as earthquakes and volcanic explosions, that take place in a matter of seconds. Observing and measuring such events requires different instruments and techniques from those used to study deep time.

Qualities of Geological Research. Geology is a field-based science. Many important advances are made in the laboratory, either by analyzing samples gathered in the field or by using computer models of geologic processes. However, the most useful interpretations of laboratory results and models are generally based on the geologist's field observations.

Geologic research is rather like detective work. The geologist pieces together bits of evidence to form a clear picture of past events. Such detective work does more than explain the past, however; it can also help the geoscientist predict what natural processes will do in the future. For example, geologists can map the locations of **faults** that moved during past earthquakes. Digging trenches across such a fault can expose layers of **sediment** that were deposited along the fault after ancient earthquakes. Geologists can then collect samples of carbon-rich material from the ancient soil just above the **bedrock** and samples of other carbon-rich material, such as charcoal formed during ancient forest fires, from

Volcanologists sample gases from the fumarole field atop Vulcano Island in Italy.

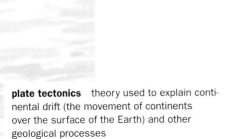

sediment above the ancient soil. Analyzing the materials for age allows scientists to bracket the date of the last earthquake using "before" (soil) and "after" (charcoal) ages. This information helps geologists predict the time of the next earthquake—not to the day or year, but within a few hundreds, thousands, or tens of thousands of years.

Recent Advances in the Earth Sciences. Since the mid-1900s, research in the earth sciences has led to major advances in knowledge. The theory of **plate tectonics** has changed human understanding of the Earth and the processes that have shaped it. The geologic record has revealed information about disasters such as impacts of the meteorites and comets on the Earth, episodes of extreme volcanic activity, and extraordinary floods and fires. Researchers are exploring possible links between these events and mass extinctions, such as the disappearance of the dinosaurs.

New technology has contributed to many advances. Satellites orbiting the Earth have given us the ability to observe and measure the oceans, continents, and atmosphere of the entire planet. Laboratory instruments capable of analyzing extremely small concentrations of material have provided more accurate dates for rock and mineral samples, revealing details of their geologic history. Faster computers with greater memory allow geoscientific researchers to create detailed, three-dimensional models of the Earth's interior or of natural processes such as **continental drift** and mountain formation.

plate tectonics theory used to explain continental drift (the movement of continents over the surface of the Earth) and other geological processes

continental drift theory that the continents move over the surface of the Earth on large segments of the Earth's crust and uppermost mantle (region just below the crust)

Areas of Research. The National Research Council (NRC) represents three private honor societies for leaders in medicine, science, and engineering. One role of the NRC is to advise the federal government on scientific, technological, and policy issues. Another is to report on the current standing and future prospects of scientific disciplines, including the earth sciences. A 1993 NRC report, *Solid-Earth Sciences and Society,* outlined many areas in which research is needed in the earth sciences.

Several of these areas involve basic research to improve the scientific community's understanding of certain Earth processes. Questions to be answered include how fluids have moved on and within the Earth to create landforms, volcanoes, and usable resources; how the Earth's crust and **mantle** formed, and how they interact to create volcanoes and earthquakes; and how the planet's deep interior, including its magnetic field, functions. Each of these and other questions includes a number of specific research topics. For example, to learn how the oceans, atmosphere, and life have developed and changed and how the Earth will respond to environmental changes in the future, scientists may study new techniques for dating soil development, past climate change as revealed by ice and sediment samples, links between the atmosphere and the ocean, and sea level changes that occurred in the past.

Other research topics in the earth sciences are directly connected to human comfort and safety. Some are aimed at sustaining sufficient supplies of natural resources, including water, energy, and minerals that are needed in everyday life. Others are geared to protecting people and property from geologic hazards, such as earthquakes and landslides. A final group of research topics focuses on keeping environmental damage to a minimum and preparing for unavoidable environmental changes at both local and global levels. SEE ALSO CARBON-14 DATING; EXTINCTIONS; GEOLOGIC TIME; IMPACT CRATERING; PLATE TECTONICS.

Earth's Crust, History of

mantle region of the Earth between the molten core and the outer crust

plate tectonics theory used to explain continental drift (the movement of continents over the surface of the Earth) and other geological processes

mantle region of the Earth between the molten core and the outer crust

sediment soils, rock particles, and other materials that are deposited over time and make up the ground, whether on dry land or at the bottom of a body of water

The young Earth was tremendously hot, perhaps partly molten, but it has been cooling steadily over geologic time. As it began to cool, a thin layer of matter hardened at its surface to form a crust. Over billions of years, that crust has been made, remade, and added to many times. Some parts of Earth's crust underlie the oceans, while other parts form the continents. The crust continues to move and change in response to powerful forces operating deep within the Earth, especially **plate tectonics.**

Nature of the Crust. The Earth's crust is a fairly thin layer that separates Earth's **mantle** from the atmosphere. The mantle beneath the crust accounts for much of the volume of the planet's rocky material. The lower (deep) mantle is hot and can move like plastic. Closer to the surface, however, it becomes colder, harder, and more brittle.

The crust and the uppermost mantle form a rigid layer, broken into enormous plates that are lighter than the underlying mantle and float on top of it. There are two kinds of crust, oceanic and continental. The crust of the ocean basins is made of basaltic and related rocks that form from material coming directly from the mantle, with layers of **sediment** on top of the rocks. The continental crust is made of granite and many other kinds of rock in complex, commonly layered arrangements. Oceanic crust averages about 7 km in thickness. Continental crust is thicker, averaging generally about 35 km, although it may be as thick as 80 km under large mountain ranges such as the Himalaya.

Formation of the Crust. Soon after the Earth formed, about 4.55 billion years ago, it developed a two-part structure. At the center was a dense, molten core rich in iron. Floating on this core was a lighter solid layer, the mantle, that was rich in silicon and magnesium, among other elements.

Heat began traveling outward from the inner core through a process called convection. Convection occurs when heat causes matter to expand and move. As it expands, it becomes less dense, and therefore it rises, carrying heat with it. For example, hot air rises towards the ceiling of a room because of convection.

In the same way, the hottest matter in the Earth's core moved to the outer rim of the core, where it caused the mantle to melt in places. Molten material from the mantle, called magma, rose through the mantle to the surface, where it cooled and hardened to form oceanic crust.

This crust was not a single, solid sheet wrapped around the Earth. Rather, rigid units of crust and uppermost mantle—known as plates—formed at the points where upwelling magma reached the surface and began to cool. Together the crust and mantle that form the plates are called the lithosphere. The thickness of the lithosphere varies from only about 10 km beneath areas where oceanic crust is forming to hundreds of kilometers beneath old continents; it averages 75 km.

Plate Tectonics and the Crust.   As the first plates began to form, they moved across the Earth's surface in a process known to geologists as plate tectonics. This process, which continues today, kept producing new crust as magma welled up through cracks in the oceanic lithosphere. The areas of up-welling were (and are) located along high ridges on the ocean floor, and the new crustal material moved away from these ridges as new igneous rock formed. Geologists call this **phenomenon** seafloor spreading.

All the while, convection continued to play a role in the Earth's great heat engine. As new hot material rose and moved out across the seafloor, old cool material was drawn down, or subducted, back into the mantle in deep trenches called subduction zones. The forces of plate tectonics have been making, un-making, and remaking the crustal plates this way for billions of years.

Continental crust began as especially thick sections of oceanic crust. Although thick, these sections were light because the melting process had removed most of the heavy, dense minerals from them. These relatively light plates were too buoyant to be recycled into the mantle at subduction zones, so they remained on the Earth's surface. Over time they collided and stuck together to form the larger landmasses called continents. Very old grains of minerals from continental rocks have been found in ancient sediments, indicating that continents began to form perhaps as long as 4.3 billion years (Ga) ago.

Early in the Earth's history, the mantle was hotter than it is now, and mantle convection was more vigorous. Oceanic plates were created and consumed

phenomenon fact or event that can be observed

Volcanologists from the Hawaii Volcano Observatory stand on the fresh crust formed over a recent lava flow. The lava is still warm enough in some places to melt their shoes.

faster, and continental collisions were frequent. As the Earth aged, it cooled, and plate production and consumption slowed down. Before 2.4 Ga ago, the average life span of an oceanic plate from creation to subduction was 20 million years. Today it is 60 million years. The rate of continental growth is not yet known for certain, but many geologists believe that more than half of the modern continental crust existed by 2.4 Ga ago.

Hot Spots, Plumes, and Super-Plumes.   The Earth's crust is a giant geological record. By analyzing rock samples and their relationships in the field, geologists can determine the depth in the mantle at which they originated and the processes that brought them to the surface. Some of the most dramatic geological processes occur above certain features of the mantle.

Regions of hot, upwelling mantle 50 to 100 km across are called hot spots. Where hot spots occur under oceanic crust, they create volcanic island chains such as the Hawaiian and Aleutian islands. Under continental crust, they can produce volcanoes on land, unless the crust is too thick or moving too fast for the hot spot to melt through and form volcanic rocks.

Mantle plumes are upwelling regions much larger than hot spots—as large as 500 to 2,000 km across. A mantle plume that upwells beneath an oceanic plate produces a large, raised plateau on the ocean basin. When a continental plate passes over a mantle plume, the plume melts the plate, causing a rift (or split) in it. The outflow of molten matter from such a rift can cover a wide area and is called flood basalt magmatism. A plume remains active for geologically brief periods of time, between 3 and 14 million years. After that its activity weakens, and the plume becomes a smaller hot spot that can persist for several hundred million years.

A period of especially rapid tectonic and crustal activity occurred between 124 and 84 million years ago. Seafloor spreading speeded up, many large oceanic plateaus formed, and volcanoes were vigorous. Geologists call this episode a "super-plume" event. They do not know exactly what caused it, but they have found evidence of several other super-plume events in the last half billion years. The Earth's crust, it seems, has been affected by periods of rapid change as well as long eras of gradual evolution. SEE ALSO CONTINENTAL CRUST; CONTINENTS, EVOLUTION OF; HEAT BUDGET OF THE EARTH; MANTLE; PLATE TECTONICS.

Echinoderms

phylum (*pl.* phyla)   second broadest classification of organisms into related groups according to their characteristics, such as mammals, reptiles, and other animals that have backbones in the Chordata phylum

radial symmetry   having similar structures in a regular pattern around a central point

Echinoderms are a diverse group of marine animals that includes such organisms as starfish, sea cucumbers, sea urchins, and sand dollars. The term *echinoderm* (meaning "spiny-skinned") refers to the fact that many of the members of the group have spines on their bodies. However, there is no one structural characteristic that is shared by all members of the **phylum** Echinodermata. In general, echinoderms have a five fold **radial symmetry** and an endoskeleton (inner skeleton) formed of many plates. Also, all modern echinoderms move and gather food by pumping water through their bodies.

Echinoderm Groupings.   There are about 15 extinct classes and five living classes of echinoderms. The classes are divided into two major groups: eleutherozoans, which move about freely in the oceans, and pelmatozoans, which fix themselves to the seafloor by means of a stalk or stem. Different classes within each group are distinguished by factors such as body shape, position of the mouth and anus, and the number and arrangement of the plates of their endoskeletons.

Echinoderms

GT   See Geologic Timescale on page viii.

Crinoids.   The largest, most diverse group of stalked echinoderms is the crinoids. Fossil crinoids are valued for their delicate beauty. The crinoids originated in the Cambrian era[GT] (570–505 million years, or Ma, ago) and flourished in the Mississippian and Pennsylvanian periods of the late Paleozoic era (360–286 Ma ago). Crinoid remains from this time were so abundant that thick limestone beds made up of their fossilized skeletons formed in the shallow seas that once covered the middle of what is now the United States. This stone, called "Indiana limestone," is a superior building material and can be found in the foundations of some of the world's most famous buildings, including the Empire State Building.

Compared to the late Paleozoic era, there are few modern stalked echinoderms. In fact, it was long thought that they were extinct since they could not be found in the shallow tropical waters they inhabited during the Paleozoic era. It was not until the 1960s that deep-sea photographs and expeditions found stalked crinoids growing deep in the ocean. Scientists believe that the evolution of bony fish made stalked crinoids easy prey in a shallow water environment, so gradually they moved to deeper water, where there are fewer predators. Modern shallow-water crinoids, called comatulid crinoids, lack the stems that would place them higher in the water and expose them to predators but also give them better access to food. Some of them climb up tall sea fans and perch on the edge, spreading their arms to catch food drifting by. Comatulids typically hide during the day and come out at night to feed.

Echinoderm Evolution.   More major groups of echinoderms are extinct than are living. During the Cambrian period, there was a wide variety of major types of echinoderms, but only a few ancestors of modern echinoderms survived the end of the Permian period, 245 Ma ago. The reason for this is unclear, but it may be characteristic of how life-forms develop. Many "experimental" designs may appear early in the history of a phylum, with the more successful ones surviving and the others dying out. However, scientists are still unsure which factors favored the survival of some types of echinoderms, but not others. SEE ALSO EXTINCTION; FOSSILIZATION AND THE FOSSIL RECORD; GEOLOGIC TIME; LIFE, EVOLUTION OF.

Eclipses and Occultations

The Moon appears as a bright body in the sky because it reflects the light of the Sun. An eclipse of the Moon, or lunar eclipse, occurs when Earth passes between the Sun and the Moon. In this case, Earth blocks the Sun's light, so it cannot reach the Moon and bounce back to Earth. A true eclipse takes place when a body disappears (or becomes much dimmer) because light has been blocked by a second body and cannot reach the first; the first body is in the second's shadow.

The Sun produces its own light by nuclear fusion. A solar eclipse takes place when the Moon passes directly between the Sun and Earth. In this case, the affected object—the Sun—is blocked from view on Earth by a solid physical presence, the Moon. The term *solar eclipse* is the popular name for this event, but it is actually an occultation of the Sun by the Moon. An occultation takes place when an object in the sky is physically blocked from view by a second body; the observer is in the second body's shadow.

Frequency and Duration.   The Moon's orbit around Earth is not parallel with Earth's orbit around the Sun. It tilts at an angle to Earth's orbit. An eclipse can take place only when the Moon's path crosses through the plane of the Earth's orbit. Only then can the Sun, Moon, and Earth be positioned in a straight line, so that one body can cast a shadow on another.

The Earth, Moon, and Sun do not often line up precisely, so eclipses are not frequent. A lunar eclipse happens as often as twice a year. A solar occultation, or solar eclipse, is visible from some part of Earth about every year and a half. In addition, an occultation is brief—usually only about three minutes long—because the Moon casts a relatively small shadow, or umbra, on Earth.

Solar and lunar eclipses. The solar eclipse shown in the upper diagram is visible over only a narrow path on Earth because of the relatively small shadow (umbra) cast by the Moon. In the bottom diagram, the Moon is totally eclipsed as it passes through the umbra (Earth's full shadow); the Moon is partially eclipsed when it passes through the penumbra, in which the Sun's light is partially blocked.

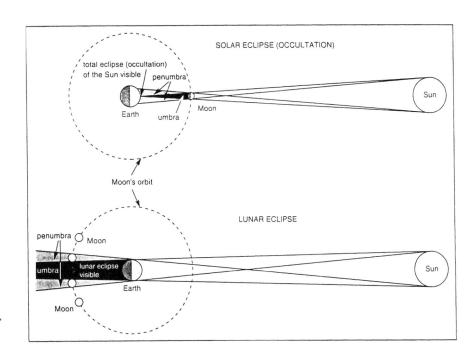

The Sun's corona, a halo of gas, surrounds the Moon during a total solar eclipse (occultation).

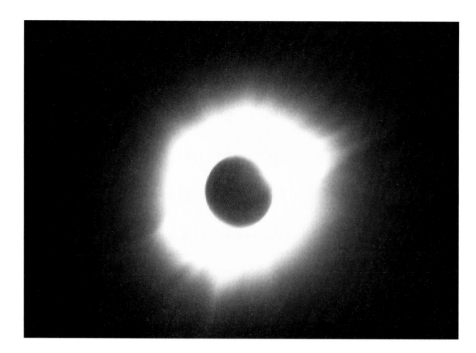

Because Earth's shadow is larger than the Moon, a lunar eclipse lasts around three hours—the whole time the Moon is passing through Earth's shadow.

Scientific Studies. A lunar eclipse is dramatic. Because the Moon still receives some sunlight filtered through Earth's atmosphere, it can look orange or blood red. An occultation, however, has more scientific value. During a solar "eclipse," the Moon blocks the Sun's bright disk (photosphere), but not its faint outer atmosphere (corona). Scientists use the opportunity to study the corona and nearby stars—which are not normally visible—through various instruments. Note that one should never look directly at the Sun during an eclipse or at any other time; harmful solar rays can damage eyesight and even cause blindness.

Eclipses and occultations can involve the other planets and their moons, or even stars, as well as the Sun or Moon. In fact, when a planet passes between Earth and a star, blocking the star from view, Earth-based telescopes reveal unique views of the planet (the "occulting body"). This sort of observation allowed astronomers to discover the rings of Uranus in 1977.

Economic Geology

Economic geology is the use of geologic information and ideas to look for and evaluate mineral deposits. These mineral deposits supply fuels, metals, and other materials for use in human activities.

Mineral Deposits Versus Valuable Ores. In scientific terms, all bodies of rock are mineral deposits. However, few contain sufficient quantities of valuable minerals (such as garnet, diamond, or quartz), metallic elements (such as aluminum, copper, gold, or iron), or nonmetallic elements (such as phosphorus or sulfur) to make them profitable. Economic geology is concerned with finding those mineral deposits that are ore bodies. An ore body is a mineral deposit that is well defined in size (three dimensions) and grade

A geologist examines the reddish soil from a good diamond-producing pit at a diamond mine. This type of soil, known as "diamond-diferous boulder grand," is common in diamond-rich areas.

(concentration of the components of high worth) and from which one or more minerals, metals, or nonmetals may be profitably extracted.

Several factors affect the value of an ore deposit. Both the size and the grade of the deposit are important. Other factors include the location of the ore, the region's climate, the deposit's depth below the surface, the shape of the ore body, the availability of power and water, any environmental considerations, the method of mining needed to collect the ore, and the technology available to turn it into a profitable material.

Searching for Ore Deposits.   Exploration for new deposits of valuable ore is based on careful study of known ore deposits, current knowledge of the geologic processes that form such ore deposits, and insights gained from research in the field and laboratory.

Economic geologists have developed various classifications of mineral deposits that reflect this understanding of the geologic environment and of processes by which different ore deposits are formed. Other classifications focus on facts such as the deposit's main mineral or element, the rock type, and the geographic locale. Much of the detail in all of these classifications is based on studies of deposits that have already been discovered and mined. The first step in such research—detailed geologic maps—serves as the basic framework for both practical and theoretical understanding of ore deposits.

The study of mineral deposits involves many areas of earth science, including **geochemistry, geophysics, mineralogy, petrology,** and **structural geology.** For example, geochemists perform chemical analyses of rocks to determine the concentration of metals, the composition of mineral **crystals,** and the sources of various ores. Geophysicists study the physical properties of rocks and ores to predict where mineral deposits may be buried or hidden. Scientists also use aerial photography and satellite images to identify surface rocks and soils that are commonly associated with ore deposits.

The Future of Economic Geology.   The search for new ores is becoming increasingly difficult because most major deposits with obvious identifying features have already been discovered. The deposits that remain undiscovered are mostly hidden deep underground or under the oceans.

A major challenge in the future will be finding replacements for the ores that are now being mined. Economic geologists will have to use the information gained from current research and apply it to new exploration strategies. They have already made impressive strides, discovering valuable ores in a number of new locations around the globe. As research continues and knowledge expands, other exploration efforts will no doubt find new sources of economically important minerals, metals, and nonmetals on Earth as well as on the Moon, the other planets, and the asteroids. SEE ALSO MINERAL DEPOSITS, EXPLORATION FOR; MINERAL DEPOSITS, FORMATION OF; MINERALS; MINERAL SUBSTANCES, USEFUL; ORE.

geochemistry   science that deals with the chemical composition of the Earth's materials

geophysics   science that deals with the physical characteristics and structure of the Earth

mineralogy   science that deals with minerals and their properties

petrology   branch of geology that deals with the origin, structure, composition, changes, and classification of rocks

structural geology   branch of geology that deals with the displacement of masses of rock

crystal   any solid whose internal structure is arranged in a repeating, orderly pattern

Ekman, Vagn Walfrid
1874–1954
Swedish oceanographer

hydrodynamics   science that deals with the forces that affect fluids (liquids or gases) and their motion

The Swedish scientist Vagn Walfrid Ekman did pioneering work in the field of oceanography. His studies of ocean currents and of the relationships among currents, surface winds, and the Earth's rotation added a great deal to the knowledge of the oceans.

Born in Stockholm, Sweden, Ekman was the youngest son of scientist F. L. Ekman, also an oceanographer. The younger Ekman received a degree in mathematics and physics from Uppsala University in 1896. After graduation he returned to Stockholm and attended lectures on **hydrodynamics** given by the Norwegian scientist Vilhelm Bjerknes.

From 1902 to 1908, Ekman worked in Oslo, Norway, for the International Council for the Exploration of the Sea. In 1910 he became professor of mechanics and physics at the University of Lund in Sweden, where he remained until his retirement in 1939.

In 1899 Ekman published a paper on the saltwater bottom currents that travel upstream in rivers as they enter the sea. Contemporary students of oceanography know that this so-called "Ekman layer" is caused by the wind's action on the surface of the ocean in combination with the Earth's rotation.

In 1905 Ekman published an expanded theory on wind-driven currents, demonstrating that the presence of coasts makes it possible for winds to influence deepwater currents indirectly. This famous theory explains ocean currents along coastlines throughout the world. Ekman's work inspired a younger generation of scientists, including Carl-Gustaf Rossby and Henry Stommel. SEE ALSO BJERKNES, V.F.K. AND J.A.B.; OCEANOGRAPHY; ROSSBY, CARL-GUSTAF A.; STOMMEL, HENRY MELSON.

Elements

See Earth Materials, Chemistry of.

El Niño and La Niña

El Niño is a global climate cycle that begins in the tropical Pacific, usually every three to seven years. The name *El Niño* (Spanish for "baby boy") reflects the appearance of the cycle's effects off the coast of Peru around Christmastime. In that region, El Niño causes increased rainfall and a decline in the fishing industry. These effects are part of a much larger climate pattern.

The atmosphere and the oceans are closely linked in the equatorial region. Under normal conditions, the trade winds blow from east to west as air rises over warm water in the west and sinks over cold water to the east. The rising air produces rainfall in the western equatorial Pacific, while the eastern Pacific region has little rain. The trade winds also push warmer surface water westward along the equator, allowing colder, nutrient-rich water to upwell, or flow up from deeper levels, to the east. Thus the winds maintain the ocean temperature difference, and the temperature difference keeps the winds blowing.

During El Niño, the normal trade winds weaken and may even reverse direction. The westward flow of ocean water at the equator then slows, causing a draining of the western warm water towards the east. Equatorial ocean upwelling is reduced, resulting in warmer sea surface temperatures in the eastern Pacific. As the western warm water cools slightly and the central and eastern equatorial Pacific warms, the ocean temperature difference decreases, the trade winds become even weaker, and El Niño grows.

As El Niño develops, the normal pattern of air rising over warm water in the western equatorial Pacific moves eastward. This pattern shift results in drought in the western Pacific, including Indonesia and Australia, and increased rainfall in the central and eastern Pacific. The warm water in the eastern Pacific spreads to the west coast of the Americas and splits to flow north and south. The normal upwelling off northern Peru weakens, drawing up only warm, nutrient-poor water. The result is a decline in production in this important region of **fisheries.** If the El Niño occurance is particularly strong, its effect in the ocean can reach as far north as the California coast.

The opposite phase of the El Niño cycle is called La Niña. La Niña is a particularly strong occurrence of other normal conditions. During La Niña, the trade winds are especially strong, and the ocean temperature difference between the western Pacific and the eastern Pacific is particularly large. Rainfall is heavy in the western Pacific but very slight in the eastern Pacific.

Effects of El Niño and La Niña from the tropical Pacific can reach far to the northeast and southeast through the atmosphere. The cycle thus affects South and Central America and, when especially strong, can have an impact on the United States. If El Niño is especially strong as occurred in 1997, unusually heavy rainfall occurs in California, with tornadoes and other heavy storms in the southeastern United States.

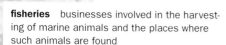

fisheries businesses involved in the harvesting of marine animals and the places where such animals are found

Major progress has been made in predicting El Niño about one year in advance because the sequence of events associated with El Niño is often the same. Detection of early signs, such as the appearance of warm water in the eastern tropical Pacific or a change in the strength of the trade winds, often allows prediction of changes in rainfall and air temperature later in the year throughout the Pacific region. An ocean atmosphere observing network and computer models now assist in studying and forecasting El Niño occurrences.

Elsasser, Walter M.
1904–1991
German geophysicist

geophysics   science that deals with the physical characteristics and structure of the Earth

plate tectonics   theory used to explain continental drift (the movement of continents over the surface of the Earth) and other geological processes

quantum mechanics   branch of physics that deals with the structure and behavior of atoms and subatomic particles

nuclear physics   branch of physics that deals with the behavior, structure, and component parts of atoms

radiation   energy emitted as particles or waves

Over the course of a 60-year career, Walter M. Elsasser laid foundations for several new fields in **geophysics**, including **plate tectonics**. He also made important contributions in **quantum mechanics**.

Born in Mannheim, Germany, Elsasser began his career in science at the University of Munich, where he studied physics. In 1924 he became involved in the emerging field of quantum mechanics and introduced important ideas about the behavior of negatively charged atomic particles called electrons.

Elsasser left Germany in 1933 and began working in **nuclear physics** at the laboratory of Frédéric Joliot in Paris. In 1936 Elsasser came to the United States, where he was introduced to geophysics at the California Institute of Technology. In the years that followed, he worked at various scientific institutions and held positions at a number of American universities, including the University of Pennsylvania, Princeton University, the University of California, and Johns Hopkins University.

During his long career in the United States, Elsasser made important contributions in many areas of geophysics, including plate tectonics, atmospheric **radiation**, and the formation of the Earth. His studies of Earth's magnetic field included pioneering work on the dynamo theory—a theory explaining the relationship between electromagnetic fields and the motion of fluids, such as the molten metal of Earth's outer core. Later discoveries led to complete acceptance of Elsasser's dynamo theory as an explanation of Earth's magnetic field. SEE ALSO EARTH, MOTIONS OF; EARTH, STRUCTURE AND COMPOSITION OF; TECHNIQUES, GEOCHEMICAL AND GEOPHYSICAL.

Energy

See Biomass Fuels; Fossil Fuels; Geothermal Energy; Hydroelectric Energy; Nuclear Energy; Solar Energy; Wind, Energy from.

Energy Use Around the World

Energy, in its various commercial forms, is vital to almost every activity in the modern world. Without it, there would be no light or heat for homes and offices; factories could not operate; farmers could not plant, harvest, or transport crops; and people could not travel to and from their jobs. Energy is so important that most governments make great efforts to ensure an ample supply of it, and they may even go to war over it.

British Thermal Unit (BTU)   amount of heat needed to raise the temperature of 1 pound of water by 1°F at 39.2°F (4°C)

Energy Use and Availability

The world consumes more than 360 quadrillion **British Thermal Units** (BTU) of commercial energy every year. Industrialized countries use a large portion of

144

this energy. For example, the United States consumes 25 percent of the world's energy, even though its population is just under 5 percent of the world's total. By contrast, about half of the world's population has no access to commercial energy. In many parts of the world, people must collect firewood to supply energy for such basic needs as cooking and heating water. The amount of energy consumed worldwide is likely to grow significantly over the next several decades as the population increases and as developing nations seek to raise the living standards of their citizens.

The world's energy resources are not evenly distributed. The industrialized countries that use most of the energy cannot supply most of these resources. Between one-half and two-thirds of all the oil discovered around the world, for instance, lies in the Middle East. Some of the largest users of petroleum have little or no reserves of their own: The United States has only 2 percent of the world's petroleum reserves, and Japan has none at all. Nations that import oil are concerned that in case of war or other political trouble, the worldwide flow of oil might be cut off, with serious economic and social consequences.

Major Sources of Energy

More than 85 percent of the world's energy is supplied by fossil fuels such as petroleum, coal, and natural gas. Almost all of the rest comes from hydroelectric dams and nuclear power plants. Solar power, wind, and other sources provide less than 1 percent of the total.

Petroleum. The largest single source of energy is petroleum, which provides about 39 percent of the world's energy needs. It is widely used because it is easy to extract from the ground and transport, and because it is very versatile. In fact, petroleum is used in making plastics and synthetic rubber as well as for many other purposes besides energy. About 1 trillion barrels of oil have been discovered around the world, enough to last well into the 2000s.

Diablo Canyon nuclear power plant near San Luis Obispo, California.

New technologies are also being used to expand the amount of oil that can be recovered. For example, oil shale and tar sands can be heated to produce petroleum. It has been estimated that these sources contain about 500 times as much petroleum as all of the world's other known sources of oil. Processing of tar sands in northern Alberta accounts for about one-fourth of Canada's crude oil production.

Coal.   Although it is the most abundant fossil fuel, coal provides only about 25 percent of the world's energy, largely because it is bulky, and therefore harder to mine and transport than petroleum or natural gas. For this reason, coal is usually burned in power plants near railroad lines or ports to minimize transportation problems. Coal is also quite dirty, and burning it releases sulfur and **greenhouse gases** that pollute the air and may contribute to global warming. The world's existing supply of coal is large enough to last for an estimated 200 years or more. The United States and the former Soviet Union have about 23 percent each of the world's reserves, but China depends more on coal for energy and produces more than any other nation.

greenhouse gas   gas in the atmosphere that traps heat and reflects it back to Earth

Natural Gas.   About 21 percent of the world's commercial energy is produced by natural gas. Natural gas burns more cleanly than petroleum or coal because it contains no sulfur, although it does release greenhouse gases when burned. Relatively cheap to produce and easy to transport, natural gas also has other commercial uses, such as in fertilizers. Natural gas is usually transported in gaseous form by pipelines, so—unlike oil—it poses little threat of spillage. Current reserves of natural gas should be plentiful enough to last for more than 60 years. Eastern Europe and the former Soviet Union have the world's largest reserves.

Hydroelectric and Nuclear Energy.   Hydroelectric dams provide about 7 percent of the world's energy, all in the form of electricity. Although they require no fuel and do not cause pollution, dams can operate only where water

Navajo Power Plant in Page, Arizona.

Energy Conservation

Conserving energy and improving energy efficiency help slow the rate at which energy is used, allowing more time to develop alternate fuels. Turning off unused lights, installing insulation in homes, driving more fuel-efficient automobiles, and installing more energy-efficient machinery in factories are all ways to cut down on energy use. Altering traditional patterns of energy use, such as by developing cars that run on electricity, natural gas, or alcohol fuels, can also help conserve energy sources that are in short supply. Adopting these strategies can reduce pollution as well as the need to import oil.

flows from a higher elevation to a lower one, and many of the best sites for dams have already been used. Dam construction can also flood large areas of land. The proposed Three Gorges hydroelectric project in China, for example, would flood enough land to displace many people from their homes.

Nuclear energy plants in 31 countries supply about 6 percent of the world's energy, also in the form of electricity. Nuclear plants use the heat released from the splitting of atoms, so they cause no pollution or greenhouse gases. However, the growth of nuclear power use has slowed because of safety concerns and rising costs. Nuclear fusion may someday provide an almost inexhaustible supply of energy from ordinary seawater, but it is not expected to be commercially feasible until 2040 at the earliest.

Renewable Energy. Solar, wind, and geothermal energy (energy drawn from hot water or from hot, dry rocks deep within the Earth) are called renewable sources because they cannot be depleted. Wind energy is probably the most promising of these sources, as engineering advances enable wind turbines to generate electricity much more efficiently than in the past. Solar energy is collected by a variety of natural and artificial devices. The most promising technologies include biomass fuels—crops that are grown to produce energy for power plants or vehicles—and photovoltaic cells, which convert sunlight directly into electricity.

The History and Future of Energy Use

Humans probably first used energy around 500,000 B.C., when they learned to use fire to cook food and provide warmth. The need for energy increased when agriculture began—around 10,000 B.C.—and people first used animal power to pull plows and carts. Wood was still the main source of energy, however, and it would remain so for most of history. As populations grew, people in parts of the Middle East, India, and China used so much wood that they stripped large areas of forest. This practice led to soil erosion and flooding, an early example of the environmental impact of energy use.

Ancient peoples also used wind, water, and some fossil fuels to supply their energy needs. Sails captured wind to propel ships, and the Persians used windmills to grind grain and pump water for irrigation. People in ancient Babylon burned bitumen, an oil-related substance that oozes from the ground, for light and cooking. Ancient Chinese households used coal for heating and cooking, and natural gas was used to provide light and heat for the palaces of emperors. The waterwheel was invented during the latter stages of the Roman Empire for grinding grain; it was later widely used in medieval Europe to provide energy for manufacturing processes. The Dutch built many windmills to pump water out of the land so that the land could be reclaimed for farming.

Fossil fuel use was limited until about the 1700s, when England began to run out of firewood and turned more to coal for energy. As the Industrial Revolution got under way in the 1700s and early 1800s, energy use (and therefore coal consumption) in Europe and the United States increased dramatically. The petroleum era began in 1859, when the first commercial oil well began operation in Titusville, Pennsylvania. The widespread use of the internal combustion engine in the 1900s opened up new uses for oil, which soon became the world's major energy source.

Scientists are exploring new sources of energy such as fuel cells—battery-like devices in which gas or liquid fuels combine chemically to produce electricity—

and solar power from space. Fuel cells cause little pollution because the fuels in them do not burn, and since they generate little waste heat, they are more efficient than fossil fuels. Scientists are also studying another process called electrolysis, in which an electric current is passed through ocean water to extract hydrogen from it. Burning hydrogen gives off only water as a by-product, so it could be a limitless, pollution-free fuel. However, the process is currently too expensive for commercial use. SEE ALSO BIOMASS FUELS; CLIMATE CHANGES; COAL; FERTILIZER; FLOODS; FOSSIL FUELS; GEOTHERMAL ENERGY; HYDROELECTRIC ENERGY; NUCLEAR ENERGY; OIL AND GAS: RESERVES, PROSPECTING FOR, RECOVERY OF; PETROLEUM; POLLUTION OF THE ATMOSPHERE; POLLUTION OF STREAMS, LAKES, AND GROUNDWATER; SOLAR ENERGY; WEATHERING AND EROSION OF THE ATMOSPHERE; WIND, ENERGY FROM.

Engineering Aspects of Earth Sciences

Scientists explore the unknown to increase human knowledge. Engineers search for new information to solve a problem. In the past, firm distinctions were made between scientists and engineers—yet history is filled with close encounters between scientific research and engineering technique.

Mineralogy.   As civilization developed, early peoples began to use newly discovered metals, to quarry stone for buildings and monuments, and to build dams and canals. These and other practical developments laid the groundwork for the earth sciences. For example, underground copper mining began at least 15,000 years ago in the Sinai Peninsula. Once mining was established, it became necessary to understand how copper deposits were formed so that miners could find new deposits. As a result, **mineralogy** grew hand in hand with mining.

mineralogy science that deals with minerals and their properties

Hydraulics and Surface Water.   Ancient peoples, in building irrigation canals, learned a great deal about water movement (hydraulics) and how to design canals that would last. Hydraulic engineers later used large laboratory models to develop ideal mathematical relationships among channel size, slope, and material for carrying a desired volume of water. The most successful of these relationships, Manning's equation, is used by earth scientists to study water flow in natural streams and rivers.

Rivers and Rocks.   Leonardo da Vinci (1452–1519), best known as the painter of the "Mona Lisa," was also a hydraulic engineer who contributed to the scientific understanding of rivers. In Italy, watching canals being cut, he noticed shells in the rock and concluded that the hills had once been below the sea. Leonardo da Vinci's studies of rivers and floods convinced him that valleys are cut by the rivers that flow through them; the **sediment** carried by the rivers moves to the sea, where it settles and in time becomes **sedimentary rock**; eventually, the sedimentary rock is uplifted again to form mountains.

sediment soils, rock particles, and other materials that are deposited over time and make up the ground, whether on dry land or at the bottom of a body of water

sedimentary rock rock formed from deposits of sediments (soils, rock particles, and other materials) over long periods of time

In the 1790s, William Smith worked as a surveyor on canal systems in England. In areas where canals were being cut, he noticed that different kinds of rock contained different fossils. These varieties of rock and fossils formed layers, and the layers fell in a certain sequence. With this observation in mind, he could look at a surface layer of rock and predict what kind of rock lay beneath it. This information helped engineers decide where to place (and not to place) canals and became the scientific foundation for **stratigraphy**— the basis of modern field work in earth sciences.

stratigraphy branch of geology that deals with strata, or layers, of rock

Landslides.   Early in the 1900s, Karl Terzaghi, a Czech civil engineer, studied how soils move and how landslides start and move down a slope. Beginning with a geologist's description of soil and other conditions, Terzaghi applied his engineering principles to those basics. Because he recognized the importance of a team effort between geologists and engineers, Terzaghi became a role model for many later engineers and earth scientists.

Planetary Sciences.   Groundbreaking advances in the engineering of space transportation of equipment and people have supported basic scientific progress in understanding the solar system. Modern engineering and earth science—disciplines once thought worlds apart—have come to interact as do the earth systems they seek to understand.

Environmental Changes

evolution   changes in groups of related organisms occurring over time

species   narrowest classification, or grouping, of organisms according to their characteristics; members of a species reproduce only with others of that group

One of the characteristics that set Earth apart from the other planets in the solar system is environmental change. Earth has a continuous history of global environmental changes since its formation 4.55 billion years ago. Earth is also the "water planet," where water in all forms (liquid, solid, and gas) is always moving through the hydrologic (water) cycle. Finally, Earth alone has a region from the upper layers of its crust to the lower layers of its atmosphere that is able to support life—a biosphere. These three factors—constant change, water, and life and the ability to support it—are continuously interdependent.

The forces and materials that cause environmental change may be natural to Earth's environment, may come to Earth from space, or may be the result of human activity. Some natural environmental changes, such as earthquakes, rainstorms, and seasons, occur quickly. Others, such as changes in sea levels or **evolution** of **species**, take place over thousands or millions of years. From outer space, forces such as solar energy and materials such as asteroids can cause planet-wide change over various timescales. Human activity, particularly since the Industrial Revolution (starting in the 1700s), has produced many environmental changes, often more quickly than natural processes could. Environmental changes affect all parts of the Earth—the lithosphere (rocky covering), the hydrosphere (liquid water), the cryosphere (solid, or frozen, water), the atmosphere, and the biosphere—as they constantly interact.

mantle   region of the Earth between the molten core and the outer crust

tectonic plate   large segment of the Earth's crust and uppermost mantle (region just below the crust) that moves as a unit over the Earth's surface, floating on a partially molten layer of rock below

Changes on and in the Lithosphere

Earth's lithosphere—its crust and uppermost **mantle**—is made up of huge, rigid plates that support the continents or make up the ocean floors. These **tectonic plates** move slowly across Earth's surface. Some plates are moving apart (for example, the North American plate and the Eurasian plate), while others collide along their boundaries or slide past one another. In some cases, the edge of one plate slips beneath another.

Tectonic plate movements are linked with many environmental changes. Earthquakes tend to take place along plate boundaries: Many of California's earthquakes take place where the North American and Pacific plates are slipping past each other. Earthquakes can trigger landslides and, in limited areas, may lift or drop the surface level quickly. Volcanoes tend to erupt along plate edges, although some occur within plates. Volcanic activity can reshape the land surface or fill the atmosphere with so much volcanic ash that it reduces the amount of sunlight reaching the surface, so that for many months or even a few years, temperatures around the planet are lowered. Undersea volcanic eruptions and undersea earthquakes cause huge tidal waves called tsunamis.

sediment soils, rock particles, and other materials that are deposited over time and make up the ground, whether on dry land or at the bottom of a body of water

Throughout much of Earth's history, tectonic plates have been on the move. The landmasses some plates support have sometimes been separated into individual continents with oceans between them, or else joined to form one supercontinent. Different arrangements of continents and oceans, mountain ranges and ocean currents, have affected global climate and the distribution and development of new species.

Other natural changes affecting Earth's surface include tornadoes, hurricanes, and other storms that produce flooding and kill life-forms. River floods can wash topsoil away from some areas and deposit **sediment** in others. Fires started by lightning or by volcanic eruption can burn forests and grasslands. Climate changes that affect moisture or temperature can have dramatic effects on plants, and changes in distribution of plant types or populations affect animals.

Changes in the Hydrosphere

Water is distributed unevenly on Earth. The oceans contain 97 percent of it; glaciers account for 2 percent; and all other water (rivers and lakes, groundwater, soil moisture, and water vapor in the air) makes up the last 1 percent. The liquid hydrosphere is always changing. Earth's ocean currents are constantly on the move, carrying energy (as heat), nutrients, sediments, and organisms from place to place. Sea levels change, rising and falling as the amount of glacier ice on the continents changes. (If all of the existing glaciers were to melt, sea level would rise by 80 m.) Life-forms have to adapt to such changes to survive; for instance, corals that live in very shallow water must grow fast enough to keep pace with a rising sea level. Changing sea levels also transform the outlines of the continents.

Surface water and groundwater change, too, largely in response to temperature and precipitation (climate) changes, changes in sea level, and movements of tectonic plates. For example, during the last ice age, there were enormous lakes in the western United States. Utah's Great Salt Lake is a small remnant of the largest of these ancient lakes.

How "Greenhouse Gases" Work

Certain gases are "selective absorbers." Shorter wavelengths of energy pass through them easily; others do not. Much of the energy from the Sun reaches Earth as light that passes through these gases in the atmosphere and reaches Earth. Some of that light becomes heat (infrared radiation). Some of the heat bounces back toward outer space—but heat has a longer wavelength than light. Greenhouse gases absorb these longer wavelengths instead of letting them escape, keeping Earth warmer than it would otherwise be. Human activity has added large amounts of greenhouse gases to the atmosphere. The ultimate result could be global warming that changes weather patterns and raises sea levels around the world.

Changes in the Cryosphere

The Earth's cryosphere is made up of glaciers; snow cover; permanently frozen ground (permafrost); and sea, lake, and river ice. Some changes in the cryosphere are seasonal. For instance, polar sea ice expands during winter, and during summer it shrinks. The top meter of permafrost in upper Canada, among other places, melts during summer. In certain temperate areas, such as the midwestern United States, winter brings ice in the forms of snow, lake or river ice, and frozen ground. Most of the more permanent cryosphere, though, is found in polar regions or on high mountains, where temperatures are low enough for ice to stay in place from year to year. These areas are where glaciers, the main part of the cryosphere, form.

Glaciers have grown and melted throughout Earth's history, as global conditions have changed. The water in glacier ice comes mainly from the oceans. During the last ice age, about 18,000 years ago, sea level was 125 m lower than it is today. Some 83 million km³ of water was locked up in ice—50 million more cubic kilometers than in modern glaciers.

Los Angeles is barely visible through the thick smog that is closely related to automobile exhaust and industrial wastes that enter the atmosphere, combined with natural conditions that keep lower layers of the air from mixing with upper layers.

precipitation water that falls to Earth in the form of rain, sleet, or snow

extinct having died out completely, leaving no direct descendants (refers to a specific grouping of organisms)

When an ice age occurs on Earth, the effects are planet-wide. In the cryosphere, glaciers grow. In the atmosphere, climates around the planet cool, climate areas move, and **precipitation** rates change. In the hydrosphere, oceans cool, and ocean current temperatures change. In the biosphere, animals and plants move out of some regions and into others; some species become **extinct** while others evolve, or develop adaptations. Until recently it was thought that there had been four such ice ages in the past 1.6 million years; newer evidence suggests that there may have been ten or more.

Modern glaciers are located primarily in Earth's polar regions, where they have a tremendous effect on Earth's systems. The great contrast between cold air above the poles and warm air over the equator helps fuel the energetic exchange of ocean and air currents that produces weather and affects climates around the globe.

Changes in the Atmosphere

The atmosphere changes in response to natural events (such as volcanic explosions), events in space (such as sunspots on the Sun), and human activities (such as the burning of Brazil's forests). Many of these changes have a direct effect on weather and/or climate.

Solar Energy and Earth's Atmosphere. The amount of energy Earth and its atmosphere receive from the Sun changes seasonally—and also over thousands of years. Seasonal variations occur because Earth tilts on its axis as it circles the sun. This tilt causes the northern hemisphere to receive more energy during half of the year—the half that includes summer; the southern hemisphere receives more energy during the other half of the year. Long-term variations occur as the amount of tilt in Earth's axis changes over time. Distance from the sun changes, too, because Earth's path around the sun is not a perfect circle but an ellipse. Changes like these can have an impact on climates all over the Earth.

The Sun itself produces varied amounts of energy, possibly due in part to sunspots, and this too affects climate. Between 500 and 150 years ago, a period of reduced sunspot activity is thought to have cooled the planet. During this period, known as the Little Ice Age, glaciers expanded in Europe and elsewhere.

Human Activity and the Atmosphere.   Earth's atmosphere is 78 percent nitrogen, 20 percent oxygen, 1 percent argon, and small, highly variable amounts of water vapor. The remaining 1 percent includes carbon dioxide (CO_2), methane (CH_4), and many other gases. Some of these gases trap the heat from sunlight that reaches Earth's surface, much as greenhouse glass holds heat inside a greenhouse. Earth is so far from the Sun that if it were not for these naturally occurring gases—especially water vapor, CO_2 and CH_4—Earth would be much cooler than it is. In fact, during ice ages, levels of CO_2 and CH_4 in the atmosphere are lower.

Since the start of the Industrial Revolution in the 1700s, amounts of CO_2 and CH_4 in the atmosphere have been increasing. Human activities such as the burning of fossil fuels (petroleum, natural gas, and coal) release these gases into the air. About one-third of this material is taken up by the ocean; the rest remains and may cause global warming.

Scientists examine ice cores taken from glaciers to learn the chemistry of Earth's atmosphere over geologic time. Ice cores indicate that modern CO_2 levels are about 30 percent higher than they have been at any time during the past 160 thousand years. They may even reach the greatest level in 3 million years.

Scientists use computer models called General Circulation Models (GCMs) to predict how greenhouse gases might affect climate. The models estimate a global average temperature increase of 1.5 to 5.3°C for each doubling of atmospheric CO_2. Some 18,000 years ago, over a period of thousands of years, a 3 to 5°C global temperature decrease occurred and produced an ice age. Scientists wonder how an equally large temperature increase over a period of just decades, rather than thousands of years, might affect the Earth. The models seem to predict decreased sea ice and snow cover at the poles; greater surface temperature increases over land than over oceans; precipitation increases at high latitudes, in midlatitudes during winter, and in monsoon areas; and, in midlatitudes, a general tendency toward drier soil or drought.

The reliability of these models' predictions remains uncertain. Several factors are not clearly understood, or have not yet been modeled well by computers—for instance, the roles played by clouds, heat that is carried by ocean water, vegetation on land, and changes in ice caps. Other factors might act to reduce temperatures—industrial sulfur emissions, forest burning, and subsonic aircraft release particulates (bits of dirt, dust, and chemicals) into the atmosphere. Particulates tend to keep sunlight from reaching the surface and may have a global cooling effect.

Still, over the past hundred years, there have been increases in average global temperature. These increases fit the pattern scientists have predicted—for instance, the largest temperature increases are in higher latitudes (toward the poles) and some continental interiors. The increases, however, have not exceeded the range of natural climate changes. Some scientists argue that human activities are not the cause.

Changes in the Biosphere

Life-forms appeared early on Earth—more than 3.5 billion years ago—and have evolved continuously. There are some 5 to 10 million (possibly as many as 100 million) distinct species on the planet. In general, the biosphere has grown more complex.

For almost 3 billion years, the planet knew only one-celled life-forms. Multi-celled organisms began to develop around 700 million years (Ma) ago. By about 480 Ma ago, fish had begun to develop, followed by early amphibians (360 Ma ago), reptiles (320 Ma ago), and the first birds (about 145 Ma ago).

New species develop in a number of ways. Geographic separation is a good example. Some sort of change—in the lithosphere, cryosphere, or hydrosphere—separates two populations of a species. Perhaps two crustal plates press together, creating a new mountain range that separates them, or a global warming between ice ages raises the sea level, leaving different groups of that species on different islands. Over time, and often quickly, the separate populations develop differently and become separate species.

Some species become extinct. In fact, most species in the animal kingdom survive only 5 to 10 Ma before they disappear. Natural changes may result in the extinction of one or more species. Catastrophic changes may produce mass extinctions. The most famous extinction event, in which the dinosaurs disappeared, may have been caused by an asteroid that struck Earth about 66 Ma ago. In any case, a very large percentage of Earth's species became extinct around that time. Soon after the event, however, a great many new species developed, so that the overall number of species on Earth increased. This same pattern—part of the ever-changing Earth—occurred at least four times before the dinosaur extinctions. SEE ALSO Biosphere; Environment and Earth Science; Glacial Ages; Glaciers and Frozen Waters; Ozone Depletion; Plate Tectonics; Volcanoes.

Environment and Earth Science

environmental engineering branch of engineering that specializes in dealing with environmental problems.

extract to separate a substance from its source

ecosystem system formed by the interaction of a group of organisms and their physical environment

Earth science is fundamentally linked to environmental issues. The Earth is a complex system, and all of its parts—from rocks and soil to human beings—interact with other parts. In order to study the materials and processes that make up the Earth, earth scientists often must examine the ways in which people interact with their geologic environment. Human activities, such as construction, mining, and agriculture, can have a profound impact on the Earth's structures and materials. At the same time, geologic occurrences—hazards, such as earthquakes, as well as vital processes, such as soil formation—have great impact on humans. Part of the earth scientist's job is to reduce the damage people inflict on the Earth, and vice versa. **Environmental engineering** is an important part of this process.

Geologic Resources.   Since the earliest days of civilization, people have used resources provided by their geologic environment—soil, rocks, minerals, water, and fuels such as coal. People have required the skills and techniques of earth science to locate and **extract** resources, from the flint used in ancient tools and weapons to the petroleum that powers modern automobiles. However, the process of extracting, storing, and using these resources can cause dramatic, and sometimes permanent, changes in the land-water **ecosystem.** Modern earth scientists and engineers must not only find and evaluate Earth's resources for their usefulness to humans but also predict the effects that humans' use of geologic resources is likely to have on the environment.

Most resources such as minerals, rocks, and fossil fuels are not renewable on a human timescale: Once they have been used, they cannot be replaced. In addition, the process of locating and extracting these resources can be quite costly and damaging to the environment. When earth scientists test and analyze deposits of these materials, they must determine how easy or difficult it

Environment and Earth Science

sediment soils, rock particles, and other materials that are deposited over time and make up the ground, whether on dry land or at the bottom of a body of water

erosion wearing away of land by wind and water

sedimentation process in which water moves suspended particles of soil, sand, or other matter and deposits them in a new location

floodplain land along a riverbank that is sometimes flooded

levee bank built along the shore of a river to hold back floods

will be to extract the resources and use them and which parts of the environment may be damaged or destroyed in the process. For example, underground mining can cause the Earth's surface to sink, while surface mining leaves gaping holes in the landscape.

One resource that is sometimes renewable is water. Surface water found in rivers and lakes can be replaced by rainfall, but when groundwater is "mined" out of a given area, it may not be restored for hundreds or even thousands of years. To manage water resources, society depends on hydrogeologists, who specialize in understanding groundwater and aquifers (the underground layers of gravel or other forms of **sediment** and rock that contain water). Hydrogeologists may be asked to locate groundwater, to assess its quality, to judge how easily it can be removed from the aquifer, or to help a town determine how long its groundwater will last. Another group of earth scientists involved with water resources is environmental engineers, who design and install the tunnels, canals, or aqueducts that distribute the water supply.

Like water, soil is an indispensable resource, but one that forms slowly. Agriculture depends on soil, and civilization depends on agriculture, so caring for soil is critical to society. Soil can be lost through **erosion,** or its fertility can be reduced if it contains too much or too little water, too much salt, or other types of chemicals. Many human activities—including farming, logging, and construction—can lead to soil damage or soil loss. Earth scientists overseeing these activities can help to prevent erosion and maintain fertility. They can also reduce the damage caused by soils. Some soils are expansive—that is, they shrink or swell depending on the amount of water they absorb. Identifying materials like these through geologic mapping can prevent billions of dollars in damage to highways, building foundations, underground pipelines, and other structures.

Geologic Hazards. While humans may damage their environment by using up or destroying its resources, nature can also cause great damage to humans. Some natural threats, such as erosion and **sedimentation,** occur gradually over a period of months or years. However, geologic hazards—including volcanoes, earthquakes, landslides, and floods—occur suddenly. Earth scientists study these hazards, determine what harm they can do, and recommend ways to limit the damage. Sometimes laws and policies, such as banning construction in areas that are likely to flood, can prevent damage from geologic hazards. In other cases, no amount of money or environmental engineering can prevent losses.

Earth scientists can do little to prevent such geologic hazards as earthquakes and volcanic eruptions. However, they can identify and map areas where these disasters are likely to occur, helping lawmakers to restrict development in these zones. In danger zones that are already developed, existing buildings can be strengthened to make them more likely to survive in case of an earthquake. Earth scientists have also developed techniques to detect early warning signs of earthquakes or volcanic activity, making it possible to remove people from an area before disaster strikes.

Landslides and floods are generally more predictable than volcanic activity or earthquakes. They are also more common and occur in many types of regions. Humans have always built homes in the lowlands and **floodplains** surrounding rivers, but the natural behavior of rivers means that they will sometimes flood. Human-made structures have changed river systems, often increasing the risk of flood damage. Engineers have attempted to control flooding by building dikes, **levees,** and dams, but such structures did little to hold back the Mississippi River flood of 1993, which caused more than $10 billion worth of damage.

Geotechnology

The new discipline of geotechnology involves the use of modern inventions in the earth sciences. Geotechnologists use computer modeling, satellite imagery, and other new technology to deal with issues such as waste management, urban planning, the use of natural resources, and the avoidance of natural hazards. For instance, a geotechnologist might use satellite images to create land use maps of a country to help its government develop a plan for using its land without abusing its environment.

Landslides occur on hillsides when soil and rock become gravitationally unstable—likely to fall downhill if disturbed. Geomorphologists identify and map areas at risk for landslides. Sometimes environmental engineers can reduce the chances of a landslide by removing some of the water from the land in a high-risk area or by putting structures in place to improve the stability of a slope. Still, losses from both landslides and floods are increasing, mostly because the world's population is growing so quickly.

Urban Areas. Crowded cities pose great challenges for earth scientists. One problem is to dispose of the great amounts of waste that are concentrated in urban areas. Geologists are involved in all aspects of landfill design: choosing a location, excavating earth material, and the expensive investigation reports required by government agencies. Other urban problems related to earth science include dealing with the increased risk of flooding and landslides, trying to prevent mineral resources from being paved over, importing large amounts of earth materials for construction, and testing the ground to make sure that it can withstand the stress placed on it by large numbers of buildings.

Environmental Policy and Management. Environmental policies are usually controversial because the issues involved are so complex. In many cases, environmental issues cross several different scientific disciplines. For example, certain Southwestern plants use large amounts of groundwater. Some hydrologists want to eliminate the plants so that the water can be made available for people to use. Wildlife biologists, on the other hand, want to save the plants because of their role in the ecosystem, which includes providing food and shelter for birds and other animals.

There are three main approaches to environmental issues. The utilitarian, or business-oriented, approach favors developing all resources to encourage economic growth. The preservation approach seeks to protect nature and natural processes in an unchanged state. The conservation approach is a middle ground that favors careful planning in all uses of nature and earth materials in order to provide the greatest good to society for the longest possible time. When these approaches come into conflict, earth scientists can help policy makers by providing background information about the issues in question. They can determine whether an environment-related action is physically possible, cost-effective, acceptable to the community, legal, and environmentally sound.

The importance of earth scientists to environmental policy depends on their ability to collect and interpret accurate data about a given ecosystem. Their training allows earth scientists to study the physical history of an area, examine similar situations in other areas, and determine whether a given action is likely to throw an ecosystem out of balance. Scientists must be able to present their information clearly and, when necessary, to defend their data and research methods publicly. SEE ALSO EARTHQUAKES; FLOODS; FOSSIL FUELS; GEOLOGY AND PUBLIC POLICY; GROUNDWATER; LANDSLIDES AND ROCKFALLS; MINING; RESOURCES, RENEWABLE AND NONRENEWABLE; SOIL DEGRADATION; SOILS, FORMATION OF AND TYPES; VOLCANOES; WASTE DISPOSAL, MUNICIPAL; WEATHERING AND EROSION.

Erosion

See Coastal Erosion and Deposition; Rivers, Erosion and Deposition by; Weathering and Erosion; Wind, Erosion and Deposition by.

Eskola, Pentti Elias

1883–1964
Finnish geologist

petrology branch of geology that deals with the origin, structure, composition, changes, and classification of rocks

metamorphic rock rock that originates from changes in the texture and mineral composition of other types of rock under conditions of extreme heat and pressure deep within the Earth

plate tectonics theory used to explain continental drift (the movement of continents over the surface of the Earth) and other geological processes

chemical equilibrium condition existing when a chemical reaction and tendencies to reverse this reaction proceed at equal rates

Pentti Elias Eskola contributed greatly to scientists' understanding of the geology of Scandinavia and Finland. A major figure in the area of **petrology**, he introduced important concepts about the structural changes that occur in rocks as a result of metamorphism—changes in pressure, temperature, and other variables.

Born on a farm in southwestern Finland, Eskola attended the University of Helsinki, where he earned a degree in chemistry in 1906 and completed a doctoral paper on **metamorphic rocks** of the Orijärvi region of Finland in 1914. In 1920 and 1921, he studied in Oslo, Norway, with Victor Moritz Goldschmidt, a noted geochemist. At about this same time, Eskola also worked at the Geophysical Laboratory in Washington, D.C., and was strongly influenced by the petrologist Norman L. Bowen. A geologist for the Finnish Geological Survey from 1922 to 1924, Eskola was Professor of Geology at the University of Helsinki from 1924 to 1953.

Eskola was the first scientist to analyze certain chemical reactions that take place as rocks change in structure. In 1920 he introduced an idea that helped link the structural compositions of metamorphic rocks to variables of pressure and temperature. In his studies, Eskola recognized that certain metamorphic rocks are formed when basaltic rocks from the crust become deeply buried. The processes that form these rocks are now interpreted in terms of **plate tectonics,** and their occurrence in certain locations around the world provides direct evidence of ancient mountain-building events.

Eskola's most lasting contribution was his description of rocks and minerals as chemical systems that tend toward a state of **chemical equilibrium.** Most later scientists who studied metamorphic rocks have treated them in terms of this concept. A number of his other concepts regarding metamorphic rocks also remain in wide use by earth scientists.

In his own lifetime, Eskola received many awards and medals from various geological societies. The Geological Society of Finland presents the Eskola Medal to scientists once every five years in his honor. SEE ALSO BOWEN, NORMAN LEVI; GOLDSCHMIDT, VICTOR MORITZ; METAMORPHIC ROCKS.

Evolution

See Live, Evolution of.

Ewing, William Maurice

1906–1974
American geophysicist

geophysics science that deals with the physical characteristics and structure of the Earth

oceanography science that deals with oceans

seismology science that deals with earthquakes

sediment soils, rock particles, and other materials that are deposited over time and make up the ground, whether on dry land or at the bottom of a body of water

William Maurice Ewing made significant contributions in several areas of **geophysics,** including **oceanography** and **seismology.** In addition to increasing scientific knowledge of the geology of the oceans, he also dramatically changed scientists' ideas about the geologic processes that shape Earth's surface and its interior. Noted for his development of new scientific instruments and the improvements he made in existing instruments and methods, Ewing is also remembered for training others who became noted scientists.

Born in Lockney, Texas, Ewing entered Rice Institute (now Rice University) in Houston at age 16. He earned degrees in both mathematics and physics. In 1929 Ewing took a job as an instructor of physics at the University of Pittsburgh, but he transferred the following year to Lehigh University in Pennsylvania.

At Lehigh, Ewing began to conduct seismological experiments on the continental shelf, the part of a continent that extends out from the coast and is submerged beneath the sea. His studies led to the discovery of a thick layer of **sediments** on the ocean floor. To aid his research, Ewing developed a scientific instrument called a precision depth recorder (PDR). Used to measure the depth of the ocean accurately, the PDR led to the discovery of huge, perfectly flat plains in some of the deepest parts of the oceans. This revelation led in turn to discoveries about deep undersea currents that carried sediments out from the continents and gouged deep submarine canyons into the continental shelf in the process.

In 1948 Ewing took a position as a professor of geophysics at Columbia University in New York. The next year, he became the founding director of Columbia's Lamont Geological Observatory, where he carried out much of his most important work. From 1955 until the 1970s, Ewing collected geophysical data on the ocean floors. He measured **magnetic fields, gravitational fields,** the thickness of sediments, and temperature changes within ocean sediments. His expeditions aboard research ships led to the collection of the largest library of sediment samples and other scientific data ever assembled in the deep oceans. Ewing's research led to the rapid establishment of a number of new geological theories, including **plate tectonics.**

Ewing also made considerable contributions to earthquake seismology, studying variations in the structure of the Earth's crust in various regions of the United States. His interest in other scientific fields led to the development of a theory about the periodic appearance and disappearance of ice ages and to research into climatic changes. SEE ALSO OCEANOGRAPHIC EXPEDITIONS; OCEANOGRAPHY; OCEANS, EVOLUTION AND STRUCTURE OF.

magnetic field region surrounding a magnet or an object carrying an electric current that affects the behavior of other magnets, magnetic materials, and electric currents

gravitational field area influenced by gravity

plate tectonics theory used to explain continental drift (the movement of continents over the surface of the Earth) and other geological processes

Extinctions

species narrowest classification, or grouping, of organisms according to their characteristics; members of a species can reproduce only with others of that group

genus (pl. genera) sixth broadest classification of organisms into related groups according to their characteristics, such as dogs and wolves in the *Canis* genus

GT See Geologic Timescale on page viii.

evolution changes in groups of related organisms occurring over time

An extinction is the final demise, or death, of a particular **species, genus,** or larger grouping of organisms that left absolutely no direct descendants. For example, the famous saber-toothed tigers, whose fossilized remains were found in the La Brea Tar Pits near Los Angeles, California, disappeared completely in the later Pleistocene epoch (1.6–0.01 million years ago). No saber-toothed cats or descendants (modified forms) of saber-toothed cats exist anywhere today.

Fossils provide scientists with evidence of extinctions. Scientists of the 1800s had enough fossil evidence to recognize some extinctions in Earth's history, but they believed that these had been caused by catastrophes that wiped out all or essentially all life during that time. Then, according to their theories, entirely new life-forms appeared. Later scientists realized that extinctions mainly occurred among portions of existing plant and animal categories. In some cases, the percentages were high—the majority of organisms were eliminated from most Earth environments. In other instances, the percentages of organisms affected were low, resulting in only minor extinctions.

Recognizing Extinction Events. Some natural events can appear to be extinctions when they are not, and scientists must take care to distinguish these events from true extinctions. For example, there is the continuous **evolution** of species and genera of organisms into adapted species and genera. Although it might appear that a group of organisms has disappeared, it may have actually evolved into a new form. Scientists focus on three aspects of the fossil record to help them identify true extinctions: first and last appearance of organisms, changes in their abundance, and changes in community structure.

Plotting the first and last appearance of an organism in the known fossil record is an approach that works well for organisms that are abundant in the fossil record, especially if the appearance/disappearance pattern occurs in widely separated locations. The procedure is inadequate, however, for rare or uncommon organisms; for soft-bodied organisms, which rarely turn into fossils;

and for organisms whose skeletons fall apart after death into a mass of unidentifiable bits.

Changes in the abundance of organisms through time can also indicate extinctions. For example, a large percentage of crossopterygian (tassel-finned) fish disappeared from the fossil record at the end of the Cretaceous period, about 66 million years ago. Although one species of these fish, a coelacanth known as *Latimeria chalumnae,* exists today in some deeper waters off East Africa, the change in crossopterygian abundance is evidence of a major extinction event.

A community is an established combination of genera and species in which the abundance of some organisms, compared with the abundance of others, remains about the same. Major changes in community structure are powerful indicators of extinction events. During every extinction, some established community organisms disappear, new groups appear, and changes occur in the abundance of those groups that remain.

Causes of Extinction. As scientists research possible causes of extinctions, they must look for explanations that allow for the extinction of some groups and the survival of others. A theory of sudden global cooling that is sometimes used to explain the many extinctions of the Cretaceous-Tertiary boundary, for example, could account for the disappearance of some four-limbed animal groups, including the dinosaurs, but it could not explain the survival of others.

Trace amounts of **iridium** at the Cretaceous-Tertiary boundary are high enough to suggest that asteroid or comet collisions with Earth at that time were responsible for the many extinction events. This possibility attracted much attention in the 1980s, but it is clear that extinction events are more complex than single cause-effect episodes.

Most of the discussions of the 1980s centered on physical changes on Earth, such as changes in climate or oceans, that the impact of asteroids or comets might have caused. However, scientists now agree that biological factors such as disease, evolutionary change, or the collapse of communities, must also be considered.

Many questions remain. Researchers continue to explore the issues of why some organisms become extinct while others survive; how long humans have played a role in the extinction of other organisms on Earth; and whether extinctions occur over a time span of a few thousand years or millions of years. SEE ALSO FOSSILIZATION PROCESSES; IMPACT CRATERING.

iridium very heavy metallic element found in meteorites (solid bodies that have fallen to Earth from space), but rare in rocks of Earth origin

Extraterrestrial Life, Search for

extraterrestrial from beyond the Earth

People have always wondered whether life existed on other planets besides Earth. Some have thought it unlikely that a single planet could be the only source of life in the vastness of the universe. The Greek philosopher Metrodorus expressed this view in the 300s B.C. when he argued: "To consider the Earth as the only populated world in infinite space is as absurd as to assert that in an entire field of millet, only one grain will grow." Modern scientists are actively searching for signs of **extraterrestrial** life, both in the Earth's solar system and elsewhere. So far, however, no one knows whether life of any sort exists outside Earth's biosphere.

Within the Solar System. Tales of civilizations on Earth's neighbors, such as Venus, Mars, and the Moon, are simply science fiction. Scientists have found no definite evidence of life on other worlds in the solar system—not even such simple life-forms as bacteria.

Humans have often imagined that life might exist on Mars. It appeared that there might be a scientific basis for such ideas after two unmanned Viking spacecraft explored Mars between 1976 and 1982. Photographs taken from orbit showed clear evidence that water had once flowed on the dry planet, a fact that raised the exciting possibility that life could have originated in the waters of Mars as it did on Earth. However, when the Viking astronauts tested Martian soil for traces of biological activity, no convincing evidence of life appeared. It remains possible that life—or the remains of long-dead life—may be found in deeper levels of soil and rock or in some other part of the planet.

Some scientists are looking on Earth for evidence of life on Mars. A number of meteorites are known to consist of material that originated on Mars. In 1996 a team of scientists announced that one such meteorite contained fossilized traces of ancient, bacteria-like life. Later, however, other experts produced evidence to the contrary. Although most now agree that the first announcement was in error, the examination of Martian meteorites continues.

Other bodies in the solar system, especially Jupiter's moon Europa and Saturn's moon Titan, may have experienced enough chemical evolution to support life, although most scientists believe that this possibility is remote. Further research is needed, but it appears that at least in the modern era, life within the solar system is limited to the Earth.

Beyond the Solar System. Another approach to the search for extraterrestrial life is to listen for radio signals from outside the solar system. Astronomer Frank Drake, who conducted the first such search in 1960, came up with a formula to estimate the number of possible civilizations in our galaxy that might be advanced enough to send out radio signals at any given time. Known as the Drake equation, this formula is based on theories about the evolution of stars and planets, chemical and biological evolution, and the development of technology and civilizations. Although parts of Drake's equation may be disputed, his basic idea remains a useful tool for scientists.

The Earth's galaxy—the Milky Way, with some 400 billion stars—is only one of hundreds of billions of galaxies in the observable universe. Astronomers have learned that there are many stars like the Sun and that the chemical elements of life are widespread in the universe. Scientific theories and astronomical evidence indicate that planets are also fairly common. Since some of those planets should contain liquid water, life may have arisen on them just as it did on Earth about 4 billion years ago.

Extraterrestrial intelligence, however, is a separate issue. Some astronomers and evolutionary biologists believe that while life may exist elsewhere, technological intelligence is a freak accident of Earth's **evolution** that is unlikely to have happened elsewhere. Many others, however, believe that evolution favors the development of intelligence because the ability to reason increases an organism's chances of survival. According to this argument, not only is it possible that life is common; it is also probable that life often develops in the direction of intelligence and technology.

A project called SETI (Search for ExtraTerrestrial Intelligence) is testing that belief by attempting to pick up radio transmissions from technological civilizations in distant planetary systems. SETI researchers expect that whatever biological form extraterrestrial life may take, intelligent creatures will use radio waves—which obey the same physical laws throughout the universe—to communicate. **NASA** once operated a SETI program but has cancelled it. Other organizations, however, are continuing the search. Among them are the

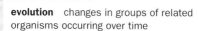

evolution changes in groups of related organisms occurring over time

NASA National Aeronautics and Space Administration, the U.S. space agency

SETI Institute, the Planetary Society, the University of California at Berkeley, and Ohio State University. If any of these research groups succeeds in detecting interstellar radio transmissions, the discovery will prove that life and intelligence exist away from Earth and that the human race is not alone in the universe. SEE ALSO MARS; METEORITES FROM THE MOON AND MARS; MOON; PLANETARY SYSTEMS, OTHER; SOLAR SYSTEM; VENUS.

Famous Controversies in Geology

Geology, the study of the Earth and the forces that have shaped it, emerged as a distinctive science only about 200 years ago. From the very beginning, geology has been marked by heated disagreements, and it has advanced to a large degree by resolving these controversies. Four major controversies highlight issues that have been important to the development of geologic thought. Three of them are concerned with one of geology's most basic questions: how the Earth came to look as it does. The fourth involves the Earth's age.

Water Versus Fire.   The visible geologic features of the Earth are varied and complex. The Earth's surface consists of many different kinds of rock. It includes plains, ocean floors, high mountain ranges with twisted and folded bands of rock, and deep gorges. Attempting to explain how these features had been formed, early geologists fell into two broad camps. One group, the neptunists, claimed that water had played the central role in shaping the Earth's surface. Other thinkers, the vulcanists and plutonists, believed instead that fire was the key to a wide range of geologic features.

mineralogist scientist who studies minerals and their properties

One of the leading neptunists was a German **mineralogist** and teacher named Abraham Gottlob Werner (1749–1817). His work reflected the widely held notion that in its earliest stages, the Earth had been wrapped in an ocean deep enough to cover even the highest peaks. According to Werner, this ancient ocean world contained chemicals that fell to the ocean's bottom and solidified to form all of the old, original rock of the Earth's surface. Newer rock came into being when mechanical forces such as erosion and weathering broke up pieces of the original rock.

Werner's theory was popular, especially in Europe, but it did not explain everything. An Italian geologist named Scipio Breislak pointed out that there was not enough water on Earth to contain all the solid material it would have taken to form the planet's crust. Werner and the neptunists could not give a satisfactory explanation of where the water had gone.

The most serious objection to neptunism concerned volcanoes. Werner argued that volcanoes had appeared on Earth only in recent times and that deep layers of a rock called basalt had been deposited by the universal ocean. However, other geologists began to find evidence that basalt is volcanic in origin, weakening Werner's theory.

New information about basalt and other volcanic rocks gave rise to plutonism, a theory that argued that volcanic fires from a heat source far beneath the Earth's surface had long played a key role in shaping the planet. The leading plutonist was Scottish geologist James Hutton (1726–1797). Those who held less extreme versions of such views were called vulcanists. Hutton's follower John Playfair (1748–1819) did such a good job of promoting plutonism that by the early 1800s, it had replaced neptunism.

Sudden Catastrophes Versus Slow Processes.   Hutton's work was also part of a related controversy in the late 1700s and 1800s—the conflict between catastrophism and uniformitarianism. This controversy had to do with the rate of geologic change.

The catastrophists believed that recognized processes such as erosion and volcanic eruptions could not account for the most dramatic features of the landscape. In the distant past, they argued, the Earth must have been subjected to massive ocean floods and titanic upheavals of the land unlike anything known in human history. Georges Cuvier (1769–1832) of France claimed that sudden, ancient catastrophes were responsible for mass **extinctions.**

extinction   dying out completely, leaving no direct descendants (refers to a specific grouping of organisms)

Some catatrophists, including Adam Sedgwick (1785–1873) of England, promoted an idea called the diluvial theory, proposing that the great flood mentioned in the Bible had created the features of the modern Earth. Geologic fieldwork, however, demonstrated that the physical history of the world is too complex to be divided simply into "before flood" and "after flood" stages.

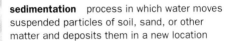

sedimentation   process in which water moves suspended particles of soil, sand, or other matter and deposits them in a new location

Hutton and others challenged the widespread catastrophist view. They did not believe that the present world was profoundly different from the world of the past, and they argued that present-day processes such as rainfall, **sedimentation,** volcanic uplift, and erosion—the wearing away of land by wind and water—could explain geologic history, given enough time in which to operate. These uniformitarians claimed that slow, gradual processes had shaped the world over long eras, and they introduced new notions of geologic time.

English scientists and thinkers of the 1820s and 1830s carried on a lively debate about catastrophism and uniformitarianism. Leading geologist Charles Lyell (1797–1875) argued strongly for uniformitarianism. His work influenced many other scientists, including biologist Charles Darwin (1809–1882), who proposed slow and gradual change in his theory of biological evolution.

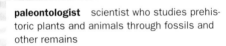

paleontologist   scientist who studies prehistoric plants and animals through fossils and other remains

The uniformitarian view came to dominate geologic thinking. In the second half of the 1900s, however, some **paleontologists** and geologists introduced new versions of catastrophism, such as the suggestion that asteroid impacts may have played a role in mass extinctions. At the same time, some biologists argued that biological evolution occurs in bursts rather than at a steady rate. The underlying ideas of catastrophism and uniformitarianism continue to influence scientific thought.

The Age of the Earth.   When geology first emerged as a science, most people believed that the Earth was only a few thousand years old. By the mid-1800s, scientists knew that it was much older, but no one knew exactly how old. The question of the Earth's age was important, however, because theories such as those of Lyell and Darwin required immensely long periods of past time.

In 1860 geologist John Phillips made the first serious attempt to measure the Earth's age. He estimated that the formation of the planet's crust had taken about 96 million years. This theory strongly challenged Lyell's ideas, which required a much longer—almost unlimited—geologic history. Another challenge came, a few years later, when the Scottish physicist William Thomson, later Lord Kelvin, calculated the Earth's age based on the time it would have taken the planet to cool to its present temperature from an original molten state. He arrived at a figure of 98 million years, but some 30 years later, he shortened his estimate to 24 million.

Critics of Kelvin challenged his assumptions and came up with alternatives that suggested a greater age for the planet. One of them, an American geologist named Thomas Chamberlin, thought that there might be energy sources, unknown to science, locked up within atoms. The discovery of radioactivity in 1896 confirmed Chamberlin's idea and undermined Kelvin's calculations. Because radioactive material in the Earth gives off energy over extremely long periods of time, the Earth has not cooled nearly as quickly as Kelvin believed.

Ancient Gods and Geologic Theories

During the late 1700s and early 1800s, all educated Europeans and Americans were familiar with the mythology and literature of the ancient world. They drew upon Roman gods for the names of three leading schools of geologic thought: neptunism, plutonism, and vulcanism. Neptunism, which held that water was the chief shaper of the Earth, took its name from Neptune, the Roman sea god. Pluto, god of the underworld, gave his name to the theory that the Earth was shaped by powerful heat sources far beneath the surface. Vulcanism, a less extreme form of plutonism, refers to volcanoes—named for Vulcan, the Roman god of the fiery forge of creation.

Scientists in the 1900s generally agreed that the Earth must be several billion years old.

Continental Drift. German scientist Alfred Wegener initiated one of the biggest geologic controversies of the 1900s when he suggested that the continents had once occupied different places on the Earth's surface and had moved into their present positions over time. Wegener was not the first to suggest that the continents had moved, but he based his theory of "continental drift," as it came to be called, on various pieces of actual evidence, such as matching geographic shapes and geological formations on opposite sides of the Atlantic Ocean.

Wegener argued that in the past, all of the Earth's landmasses had been joined together in a single large continent, which had broken apart. Slowly the continents moved into their present places; oceans formed where they had pulled apart; and mountain ranges rose where they collided. The chief problem with Wegener's idea was that he was unable to explain just how the continents might have moved across the Earth's crust.

Some scientists ridiculed Wegener's idea. However, midcentury research in ocean floor mapping and in dating of rock samples supported Wegener's belief. The theory of plate tectonics emerged in the 1960s to explain the movement of landmasses, and geologists began to piece together the history of Wegener's wandering continents. SEE ALSO CUVIER, GEORGES; DARWIN, CHARLES ROBERT; EXTINCTIONS; GEOLOGIC TIME; GEOLOGY, HISTORY OF; HUTTON, JAMES; LIFE, EVOLUTION OF; LYELL, CHARLES; PLAYFAIR, JOHN; SEDGWICK, ADAM; WEGENER, ALFRED LOTHAR; WERNER, ABRAHAM GOTTLOB.

Faults

See Structural Geology.

Fertilizer

nutrient substance that promotes growth and health in living things

Fertilizer is a substance added to the soil to maintain or increase its ability to support plant growth. In a completely natural environment, **nutrients** taken from the soil by growing plants return to the soil in the form of dead leaves, stems, or other plant debris. When plants are harvested for food, animal feed, or timber, however, the soil loses these essential nutrients. The three most important depleted nutrients—potassium, nitrogen, and phosphorous—are the major components of fertilizers. Farmers use chemical fertilizers to return these nutrients to the soil. Other important soil additives include sulfur and lime.

Sources of Potassium, Nitrogen, and Phosphorus. Farmers must reapply potassium to their fields after every harvest because plants absorb 90 percent of the potassium added to the soil. Potassium is sometimes called potash, which is actually a substance that includes minerals containing potassium or potassium in a form that can be dissolved in water and applied to crops. Mineral deposits formed during the final stages of evaporation of seawater are a major source of potassium. Other sources include granite meal, greensand (a greenish sand primarily made up of glaucomite), kelp, wood ashes, and liquid seawater. Israel, for example, has the technology to remove potassium and other elements from seawater.

The major source of nitrogen is Earth's atmosphere, which is 78 percent nitrogen. Manufacturers remove nitrogen from the atmosphere through a process in which nitrogen reacts with hydrogen to produce ammonia. This nitrogen compound is then used in the production of fertilizer. Russia and China are the world's two largest producers of ammonia, followed by the United States. Nitrogen is rarely found in minerals on the land. A major exception, however, are the deposits formed by evaporation of lake waters, such as those in the Atacama desert of Chile. Though large, these deposits cannot meet the worldwide demand.

Accumulations of bird and bat droppings, called guano, were once a large source of phosphorus and still provide about 8 percent of the world's supply. However, the major source of phosphorus is the mineral apatite. The largest reserves of apatite are found in shale rock formed in coastal ocean waters beneath areas where **plankton** were very active. Processors first separate the apatite from other minerals, treat it with sulfuric acid to form phosphoric acid, and use this acid to make ammonium phosphate, which can be used by plants.

plankton   plant and animal life, usually minute, that floats or swims at or near the surface of water

Sources of Sulfur and Lime.   A number of plants thrive in acidic soil, and sulfur adds acid to soil. Sulfur is one of the few elements that occurs naturally and can be found in a number of sources. It occurs most often among layered salt deposits—evaporites—left by evaporating seawater. Rocks, volcanic hot springs, natural gas, some crude oils, and certain metallic minerals also contain sulfur. In some places, sulfur-bearing rock is removed through **open-pit mining.** In other places, hot water is injected into the rock to melt the sulfur and remove it directly. Sulfur has many uses, but the main product is sulfuric acid, from which phosphate fertilizer is made.

open-pit mining   method of mining that scoops out mounds of earth and ore (metal-bearing mineral or rock) instead of drilling into the ground

Used as a soil additive, lime (CaO) has an effect opposite to that of sulfur. That is, lime reduces soil acidity in areas where the acid content of the soil is too high for most plants to grow. Producers form lime by heating limestone (CaO_3) or certain related minerals to temperatures of 1,000 to 1,300° C. All developed nations and most developing nations have facilities for producing lime, so the supply is virtually endless. Lime has many uses. Of all the lime produced in the United States, the largest amount is used by chemical and steel manufacturers. Construction industries use sizable amounts, and most of the rest goes into water purification and other environmental uses. Only about 1 percent is used for fertilizers. SEE ALSO MINING.

Fish Evolution

vertebrate   any animal that has a backbone or spinal column

evolution   changes in groups of related organisms occurring over time

GT   See Geologic Timescale on page viii.

dentine   dense, hard mineralized tissue of which teeth are primarily composed

Fish were the earliest **vertebrates** to appear on Earth. Their **evolution** is of special interest to scientists because a group of bony fish is believed to be the forerunner of the first amphibians—animals that can live on land or in water. These early amphibians led to the development of all land vertebrates, including humans.

Early, Jawless Fishes.   The oldest recognized fish fossils were uncovered in Australia in 1992. These fossils include scales and bone fragments of primitive jawless fishes, or agnathans, from early in the Ordovician period[GT], about 480 million years ago. These simple fishes grew to lengths of up to 20 cm and had up to 15 pairs of gill openings, tails covered by long scales, and bony armor. One group, called Osteostraci, had a thick bony shield covering the head and a tail covered in thick, platelike scales. This group is regarded as the closest relative of the jawed fishes.

Distinct from other early jawless fishes were two groups that had no armor. One group, the Anaspida, had a thin, eel-like body, and was probably the ancestor of the present-day lamprey. The other group, Thelodonti, was covered in scales with pointed crowns made of **dentine.** These scales are well preserved in

Fish Evolution

sediment samples and enable geologists to date layers of rock from the Silurian (438 to 408 million years ago) and Devonian (408 to 360 million years ago) periods when these fish lived. Among all the early jawless fishes, only the lamprey and the hagfish survive today. The others largely disappeared by the middle to the end of the Devonian period.

Jawed Fishes. Fishes with jaws and teeth (gnathostomes) existed by the beginning of the Silurian period. A number of diverse groups, with or without bony armor, came and went over the following 200 million years. One armored group, the placoderms, dominated Devonian seas and rivers. Some placoderms reached sizes as great as 10 m. Placoderms are believed to be closely related to sharks.

Sharks are cartilaginous, which means that their skeletons are made of clear, flexible tissue called cartilage, rather than hard bone. The earliest shark fossils come from the early Devonian period. About 30 different kinds of sharks are known from teeth and fin fossils from the late Devonian. Modern sharks belong to a group that appeared in the middle Devonian. The largest predatory shark, *Carcharocles megalodon,* grew as large as 15.9 m, with teeth up to 15 cm high, during the Miocene (24 to 5 million years ago) and Pliocene (5 to 2 million years ago) epochs.

Bony Fishes. Osteichthyes, or bony fishes, began to flourish from around the start of the Devonian period. They are the most successful of all living fishes, with more than 23,000 modern species of teleost—the most advanced group of bony fishes. Early bony fishes fall into two groups: ray-finned and lobe-finned.

From fewer than 10 known species in the Devonian, more than 47 **families** of ray-finned fishes had evolved by the end of the Paleozoic era, about 245 million years ago. Teleosteans began to appear by the late Triassic period (245 to 208 million years ago). These early teleosteans had a number of advanced traits, including a complex skeleton in the tail fin and a mouth whose

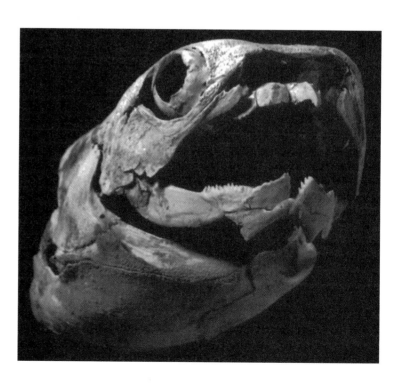

These fossil remains of a 370-million-year-old placoderm fish were found in the Gogo Formation in Western Australia.

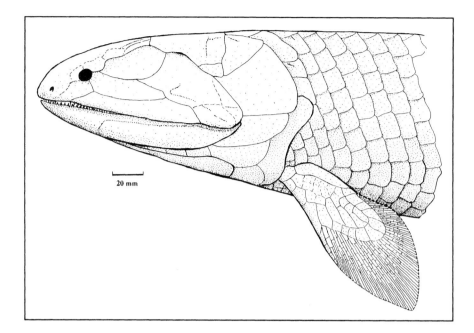

Koharalepis jarviki is a crossopterygian fossil fish from the Devonian period in what is now Antarctica.

jaws were freed from the rest of the cheek—a unique characteristic that allowed the wide range of feeding techniques that we see among fishes today.

The lobe-finned fishes, which separated into two groups, took a different route. One group, the Dipnoi (lungfishes), had both lungs and gills, so that they could take oxygen from air as well as from water. Lungfishes became highly diversified during the Devonian period, and by the Permian period (286 to 245 million years ago) had evolved the ability to lie buried in mud over dry summers to await the next wet season.

The other group, the Crossopterygii (tassel-finned fishes), had a divided brain case and jaws with fangs. This group left behind numerous fossils, and in 1938 a living species—the coelacanth *Latimeria chalumnae,* was discovered. Other specimens are still being discovered. Another tassel-finned group, the osteolepiforms, are believed to be the ancestors of the amphibians and all their land vertebrate descendants. SEE ALSO AMPHIBIANS AND REPTILES.

Floods

When the amount of water in a river is too great for its channel, a flood occurs. Water spills over the riverbanks and across the floodplain—the surrounding flat area. Floods are natural events that occur under certain meteorological conditions. They transport sediment and help maintain wetlands. From a human point of view, however, floods are among the most destructive of disasters. Researchers study floods and records of past floods to learn more about the causes and the effects of these events. One goal of such research is improved protection from the dangers associated with flooding.

Causes of Floods.   Unusual weather causes the largest floods. One example is the Big Thompson Canyon flood of 1973 in Colorado. An extreme thunderstorm dropped 250 mm (almost 10 inches) of rain on a single small area in a few hours. The rain produced a flash flood that raced down a canyon, killing 139 people and costing $35 million in property damage.

Some floods involve larger areas. In 1955 two hurricanes hit southern New England in nine days, dumping 500 mm of rain. The resulting floods were the region's worst since the 1600s. More than 200 people died.

Floods

Heavier-than-usual rainfall over a broad area for more than half a year produced flooding of the Mississippi River in 1973 and again in 1993. The rainfall saturated the soil until it could hold no more water. Flood control reservoirs were filled, and the Mississippi and its tributaries (streams that feed into the river) filled and overflowed their channels. With nowhere else to go, water pooled across vast, low-lying flatlands. The 1993 floods caused an estimated $10 billion in damage.

Predicting Floods.   To estimate the risk of flooding, scientists use a formula to calculate the likelihood that a flood of a given size will occur in a given year on a given river. The basis for the calculation is the flood history of each river, based on the historical record. The longer and more accurate the record is, the more correct the risk estimate determined by river scientists and managers is likely to be.

hydrologist   scientist who studies water movement patterns

The risk of a flood of a given size is often expressed as a time interval called the return period. For example, a **hydrologist** might refer to a "5-year flood." This does not mean that such a flood will happen every 5 years; it means that there is 1 chance in 5 (or a 20 percent chance) of such a flood happening in any given year. A 5-year flood would not be a dramatically large one. It would be a fairly common occurrence on that river. A 100-year or 500-year flood, on the other hand, would be much larger and far more rare. In any given year, people living along that river would have 1 chance in 100 (a 1 percent chance) of experiencing a 100-year flood.

Flood risk calculations are useful, but they are only estimates, not exact predictions. They are especially unreliable for estimating the likelihood of big floods. The biggest flood in a particular river's historical record is probably not the biggest that has ever occurred or ever will occur there. In some cases, researchers can use the geological record of flood deposits in an area to build a more complete flood history. The geological study of long-past floods is called paleohydrology.

Changes in climate can make flood risk hard to estimate. So can human activities, including changes in land use. Clearing forests, for example, changes water runoff patterns, and an increase in paved areas means that there is less soil to absorb rainfall.

Protection from Flood Dangers.   There are two approaches to protecting people and property from floods. One approach involves building structures to prevent rivers from overflowing their banks. The other concerns land use planning and management.

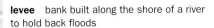

levee   bank built along the shore of a river to hold back floods

Structures for flood protection include dams to increase the amount of water that can be stored within a river basin, **levees** to allow rivers to carry more water, and channelization projects designed to speed the passage of floods through the river system. Although such projects are often effective, floods such as those in the Mississippi River basin in 1993 are reminders that no engineered system can provide more than limited protection from the risk of flooding. Dams and other engineering solutions may even give the people who live near them a false sense of confidence, and when these structures fail, the results can be disastrous.

The other approach to flood protection is to limit development in high-risk areas. To be eligible for national flood insurance, cities and towns are required to have zoning programs to regulate building within the 100-year floodplain of their rivers. Other government agencies have created zoning programs based on the elevation of a river's largest recorded flood. Hydrologists point out, however, that the historical record is too short to include the entire range of flooding possibilities. No one can say just how big the biggest possible flood could be.

Volume 1 Index

Index

Index

Index

Index

Index

Index

Radiometric dating
    archaeological geoscience, 13–14
    Carbon–14, 35–36
    cave formation, 59
    defined, 13
Radio signals, distant planetary, 159
Radio waves, 36
Radon, 122, 129
Rahonavis ostromi, 30
Rainfall. *See* Precipitation
Rain forest, climate, 67
Ray-finned fish, 164
Red Sea, 81
Redshift, 86
Refracting telescopes, 21
Refractory, defined, 104
Refractory materials, 61
Relativity, Einstein's theories of, 85
Remote sensors, 114–115, 116
Reptiles. *See* Amphibians and reptiles
Reptiliomorpha, 7
Research. *See* Earth sciences information;
  Earth sciences, research in
Resinite, 69
Resources, renewable and nonrenewable,
  1.153; *See also specific resources*
Reverse osmosis, 95
Revolution of Earth. *See* Earth, motions of
Rhynchocephalia, 7
Richter, Charles F., 126
Richter scale, 125–126
Rifting of the crust
    continental margins, *80,* 81
    continental plate, 81, 109
    magmatism, 84, 137
    oceanic crust, 109–110
Rift valleys, formation, *80,* 81
Riprap, 77
Rivers
    brackish water, 94
    floods, 150, 165–166
    hydraulic engineering, 148
    hydrosphere, 28, 65, 110
Rock paintings, 57
Rocks
    beach composition, 72, 73
    bedrock, 57, 58, 59, 108
    carbon dioxide, 123
    chemical analysis, 119
    as construction material, 76–77
    continental crust, 78, 84, 85, 108,
     109
    cycle, 122
    magnetization, 107
    mantle material, 112, 113
    materials manufactured from, 77–78
    metamorphic, 108, 156
    oldest fragments, 108
    origin controversy, 90
    oxygen traces, 24
    slippage, 124
    soft abrasives, 1

from space, 18, 50, 119
tectonic stress, 124
terranes, 82
    See also Igneous rocks; Lithosphere;
     Sediments and sedimentary rocks;
     Weathering and erosion
Rossby, Carl-Gustaf A., 143
Rotation of Earth. *See* Earth, motions of
Rouge (powdered iron oxide), 1
Royal Astronomical Society, 86
Royal Society of London, 35
Rubidium, 122

S

Saber-toothed tigers, 157
Salamanders, 6, 7
Saline water. *See* Desalination; Seawater,
  characteristics of
Salts, 110
San Andreas Fault, 82, 124
    drill hole research, 101
    major earthquakes, 128
Sand
    as abrasive, 1
    beach, 72, 73
    as construction material component,
     77
    continental crust, 108
    windblown, 26
Sand dollars, 137
Sandstone, 76, 77
San Francisco earthquakes, 128
Satellites
    asteroid, 17
    atmospheres, 24
    Earth-Moon system, 104
    See also Europa; Moon; Titan
Satellites, artificial
    archaeological geoscience, 13
    astronomical observation, 21
    Cosmic Background Explorer, 88
    debris from, 93
    remote sensing, 48
Saturn. *See* Jupiter and Saturn
Sauropodomorphs, 97-98
Sauropods, 97-98
Savanna tropical climate, 67
Science fiction, 18
Science writing
    Carson, Rachel, 56–57
    earth science careers, 55–56
Scientific creationism, 89, 90
Scorpions, 16
Scuba diving, 47
Sea Around Us, The (Carson), 57
Sea cucumbers, 137
Seafloor
    continental rise, 80, 82
    drilling research, 100
    echinoderms, 137
    hydrothermal vents, 29

passive margins, 81
plume magmatism, 83–84
    See also Marine life; Oceanic crust
Seafloor spreading, 81, 82, 83
    confirmation of theory, 100
    process, 136, 137
Sea of Japan, 84
Sea level
    erosion toward, 79
    rise in, 116–117, 150
Seals, 8
Seasons
    climate changes, 65
    Earth's motion, 102–103, 151
Sea trenches. *See* Deep-sea trenches
Sea urchins, 137, *138*
Seawalls, 73
Seawater, characteristics of, 110
    desalination, 94-95
    TDS level, 94
Secondary schools, earth science teaching
  careers, 52–53
Sedgwick, Adam, 91, 161
Sediment, defined, 12, 28, 35, 59, 68, 80,
  83, 108, 122, 133, 135, 148, 150, 154,
  156, 164
Sedimentary rocks
    defined, 102, 118, 148
    Karroo system, 102
Sedimentation, defined, 154, 161
Sedimentology
    archaeological geoscience, 13, 14
    defined, 13
Sediments and sedimentary rocks
    basin collection, 81
    bedrock upper layers, 108
    Brongniart-Cuvier study, 35
    coal formation, 67, 68
    continental rise, 80
    flood depositions, 150, 165
    mountain uplift, 148
    paleomagnetic dating, 59
Seeds
    angiosperms, 10
    atmospheric, 28
Seismic waves, 124–126
    amplitude measurement, 125–126
    defined, 108
    density shift (Moho), 100, 108
    to probe Earth's internal structure,
     106, 108, 111, 133
Seismograph, defined, 125
Seismological, defined, 90, 130
Seismologists
    defined, 113, 115, 124
    earthquake prediction, 129
Seismology
    defined, 156
    discoveries in Earth's interior, 106,
     107, 113
    earthquakes, 124–129
    Ewing, William Maurice, 156–157

Index

Super-plumes, 137
Surface water
    defined, 45
    hydraulics, 148
Synapsida, 7
Synthetic diamonds, 1
Syrtis Major (Mars), 21

T

Talc, 61
Tangshan earthquake, 129
Tar, 77
Tar sands, 146
Tassel-finned fish, 165
TDS. *See* Total dissolved solids
Teaching careers, earth sciences, 52–54
Tectonic, defined, 13
Tectonic plates
    defined, 80, 82, 108, 126, 133, 149
    earthquake occurence, 126–127
    environmental changes, 150
Tectonism, active
    continental and oceanic crusts
      changes, 82, 109–110
    continental margins, 80–82
    continental plate collision, 78–79, 82,
      84
    earthquake locations, 126–127
Teleosteans, 164–165
Telescopes, *20*, 21–22, *22*, 36
Television, earth sciences reporting, 56
Temperature
    climate extremes and records, 67
    climatology, 64
    cryosphere sensitivity, 65
    Earth's atmosphere vertical distribu-
      tion, *23*, 111
    Earth's atmosphere vs. surface, 23
    Earth's interior, 112, 114
    global cooling, 118, 152
    global surface air trends, 66
    global warming causes and impact,
      64, 116–118, 152
    hydrosphere changes, 150
    mantle reactions, 112, 113
    ocean density effects, 110
    planetary atmospheres, 24
    See also Global warming
Terranes, 82
Tertiary period, extinctions, 158
Terzaghi, Karl, 148–149
Thecodonts, 97
Thelodonti, 163–164
Thermodynamics, 32
    defined, 88
Theropods, 97, 98, 99
Thomson, William. *See* Kelvin, Lord
Thunderstorms
    convection, 25
    flash floods, 165
Ticks, 16

Tidal wave, 28
Tides, 102
    sea level changes, 117
Tiles, ceramic, 61
Titan, 159
Titanium, 72, 122
Tokyo earthquake, *127*, 128
Topography
    active margins underwater, 82
    caves and karst, 57–60
    defined, 82
Tornadoes, 143, 150
Total dissolved solids, 93, 94
Trade winds, 143
Trees-down theory (bird flight), 30
Trench accretion. *See* Deep-sea trenches
Triassic period
    fish, 164–165
    reptiles, 8
Tripoli (soft abrasive), 1, 2
Trojans (asteroids), 16
Tropical climates, 67
Tropopause, 22–23, 111
Troposphere, 22, 28, 111
Tsunami, 128
Tuataras, 7
Tuna, 8
Tundra climate, 67
Tungsten, 3, 4
Turtles, first fossils, 7, 8
Tyrannosauridae, 97
Tyrannosaurus rex, 97

U

Ultramafic rocks, 112
Ultraviolet radiation, 23, 36, 111
Underclays, 60–61
Under the Sea-Wind (Carson), 57
Underwater exploration. *See*
    Oceanography
Uniformitarianism, catastrophism vs.,
    160–161
U.S. Air Force, 18
U.S. Department of Defense, 45
U.S. Department of Energy, 45
U.S. Geological Survey, 39, 46, 91
    Astrogeology branch, 50
    World Wide Web home page, 130
Universe
    critical density, 86–87
    expanding, 86–87, 88
    galaxies, 86
    inflationary, 88
Universe, origins and evolution of. *See*
    Cosmology
Universities. *See* Colleges and universities;
    specific institutions
University of Arizona, 18
University of California, Berkeley, 5, 128,
    160
University of California, Los Angeles, 33

University of Chicago, 30
University of Helsinki, 156
Uranium, 59
Uranus
    auroras, 26
    Oort cloud, 75
    rings discovery, 140
Urban problems, 155

V

Valley formation, 57
Vegetation, climate types, 67
Veil Nebula, 87
Velociraptor, 30
Veneer, defined, 76
Venus
    astronomical observation, 21
    atmosphere, 25
Vertebrates
    comparative anatomy, 88–89
    defined, 2, 30, 89, 163
    fish as earliest, 163
    paleontology, 89
Viking spacecraft, 159
Vitrain, 69
Volatile, defined, 104, 122, 123
Volcanic ash
    climate changes, 63
    geological dating, 13
Volcanic islands, 137
Volcanic rock, 72
Volcanoes
    climate changes, 63
    continental crust composition, 84, 85
    drill hole research, 99
    as force in Earth's surface develop-
      ment, 160
    formation evolution, 90
    hazard potential, 154
    oceanic crust, 83
    oceanic ridges, 82
    as study tool for Earth's internal
      structure, 111
    super-plume event, 137
    Vostok, lowest recorded tempera-
      ture, 67
Vulcanists, 160

W

Wallboards, 78
Warm-blooded animals, 31, 98–99
Washability, coal, 70
Waste disposal, brackish water, 94
Waste products, biomass fuels, 26, 27
Water
    cave erosion, 57, 58, 59
    chemical composition, 110
    comet contents, 75, 76
    hydraulics, 148